Outage

Investment Shortfalls in the Power Sector in Eastern Europe and Central Asia

Outage

Investment Shortfalls in the Power Sector in Eastern Europe and Central Asia

Ani Balabanyan
Edon Vrenezi
Lauren Pierce
Denzel Hankinson

THE WORLD BANK
Washington, D.C.

ISBN: 978-0-8213-8738-2
eISBN: 978-0-8213-8739-9
DOI: 10.1596/978-0-8213-8738-2

Library of Congress Cataloging-in-Publication Data

Outage : investment shortfalls in the power sector in Eastern Europe and Central Asia / Ani Balabanyan ... [et al.].
  p. cm. -- (Directions in development)

Includes bibliographical references.
ISBN 978-0-8213-8738-2 (alk. paper) -- ISBN 978-0-8213-8739-9
1. Power resources--Europe, Eastern--Finance. 2. Power resources--Asia, Central--Finance. 3. Privatization--Europe, Eastern. 4. Privatization--Asia, Central. 5. Global Financial Crisis, 2008-2009. I. Balabanyan, Ani, 1978-
HD9502.E8522O98 2011
332.67'22--dc22 2011006397

Contents

Figures

Tables

Foreword

The global financial crisis severely affected economies in Eastern Europe and Central Asia (ECA). Industrial production plummeted, leading to higher unemployment and lower gross domestic product (GDP). Currencies depreciated across the region. Government tax revenues declined sharply, leading to high budget deficits and rising levels of public debt. A tightening credit supply and deteriorating financial conditions have limited the ability to borrow in the public and private sector.

For the power sector in ECA, the global financial crisis offered both a reprieve and a warning. A major investment gap existed before the crisis, as power sector companies struggled to mobilize financing for an increasing amount of under-maintained, Soviet-era infrastructure in disrepair or reaching the end of its useful life. The financial crisis slowed demand enough to delay an imminent energy shortage by a few years. In this sense, the financial crisis bought ECA countries some time. However, the same factors that slowed demand have further limited the funds public and private electricity companies have for new investment and restricted the supply of financing. An energy crisis has been postponed, but not avoided.

This report analyzes the impacts of the global financial crisis on power sectors in five countries in the ECA region: Armenia, the Kyrgyz Republic,

Romania, Serbia, and Ukraine. It estimates the investment gap and proposes a prioritization of critical investments in each country. The report also proposes actions needed to mobilize financing for the sector, including a continued commitment to legal, regulatory, and policy reform in the sector. The global financial crisis has created a window of opportunity to meet investment needs and avert a potential power shortage, but governments need to recognize and act on this opportunity. This report serves as a starting point to facilitate further World Bank engagement in the region that can help governments make timely, critical investments and foster sustainable investment in the sector over the long term.

Philippe Le Houerou
Vice President
Europe and Central Asia Region

Acknowledgments

This report has been prepared by Ani Balabanyan, Edon Vrenezi, Lauren Pierce, and Denzel Hankinson under the supervision and guidance of Indermit Gill and Ranjit Lamech. Valuable comments were received from peer reviewers Istvan Dobozi and Maria Vagliasindi. The report also benefited from valuable ideas, opinions, and expertise of Gary Stuggins, Martin Raiser, Loup Brefort, Sunil Khosla, Kari Nyman, and Gevorg Sargsyan. The report benefited from support provided during in-country visits and useful feedback on country-specific analysis from Dmytro Glazkov, Doina Visa, Mirlan Aldayarov, Miroslav Frick, and Arthur Kochnakyan. The report also benefited from detailed country reports and feedback provided by local consultants, including Eduard Nersisyan, Lilit Melikyan, Oleksii Romanov, Mirgul Aydarova, Ana Nutu, Slobodan Ruzic, and Nikola Nikolic. The authors wish to thank the numerous individuals from other international financial institutions, commercial banks, power sector companies, and government entities for providing insight and knowledge on the impacts of the financial crisis in each of the case study countries.

The report is being published by the Public-Private Infrastructure Advisory Facility (PPIAF) and the Energy Sector Management Assistance Program (ESMAP).

PPIAF is a multidonor technical assistance facility that focuses on help-ing developing countries improve the quality of their infrastructure through private sector involvement. For more information on the facility, go to www.PPIAF.org.

ESMAP, a global knowledge and technical assistance partnership administered by the World Bank and sponsored by official bilateral donors, assists low- and middle-income countries, its clients, to provide modern energy services for poverty reduction and environmentally sus-tainable economic development. It is governed and funded by a consulta-tive group comprising official bilateral donors and multilateral institutions representing Australia, Austria, Canada, Denmark, Finland, France, Germany, Iceland, the Netherlands, Norway, Sweden, the United Kingdom, and the World Bank Group.

Abbreviations

CAPEX	capital expenditures
CHP	Combined Heat and Power Plant
DSCR	debt service coverage ratio
EBRD	European Bank for Reconstruction and Development
ECA	Europe and Central Asia
EE	energy efficiency
EIB	European Investment Bank
EUR	euro (currency)
GDP	gross domestic product
HPP	hydropower plant
IFC	International Finance Corporation
IFI	International Financial Institution
kV	kilovolt
kWh	kilowatt hour
LCDP	Least Cost Development Plan
MW	megawatt
NPP	nuclear power plant
OPEX	operating expenditure
PCG	Partial Credit Guarantee
PPP	Public Private Partnership

PRG	Partial Risk Guarantee
RE	renewable energy
SHPP	small hydropower plant
SPA	share purchase agreement
TPP	thermal power plant
VAT	value added tax
WB	World Bank
WPP	wind power plant

For abbreviations used for specific power sector entities in each of the case study countries, see table A.2 in Appendix A.

CHAPTER 1

Executive Summary

Before the onset of the global financial crisis in late 2008, countries in Eastern Europe and Central Asia (ECA) experienced strong economic growth. Demand for electricity increased steadily with gross domestic product (GDP). GDP grew, on average, 6.5 percent between 2000 and 2007, and electricity consumption per capita grew 2.75 percent. Meanwhile, energy security and supply reliability were a growing concern for policymakers and planners. Despite increased access to financing through the opening of international financial markets, under-maintenance of old Soviet-era power sector infrastructure created a backlog of critical investments threatening the stability of the sector. As a result, a gap between demand and available supply capacity was beginning to emerge.

The global financial crisis hit economies in the ECA region harder than any other region (see figure 1.1). The sharp drops in GDP reduced government tax revenues, leading to rising budget deficits and higher levels of public debt.

This report analyzes the impacts of the financial crisis on power sectors in the ECA region through the experience of five countries (the study countries)—Armenia, the Kyrgyz Republic, Romania, Serbia, and Ukraine. The report's objective is to help policymakers in the region

Figure 1.1:   Gross Domestic Product Annual Growth by Region, 2000–2009

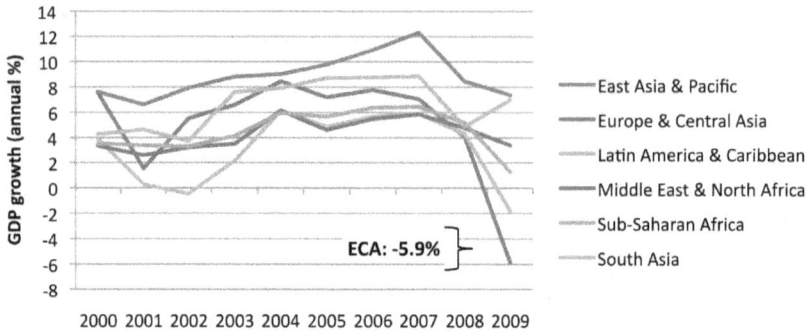

Source: World Bank 2010. Data retrieved August 28, 2010, from World Development Indicators (WDI) online data-base. Washington, DC: World Bank.

plan and prioritize electricity sector investments in the wake of the financial crisis, and to provide a basis for future discussions about World Bank assistance.

Short-Term Impacts of the Financial Crisis

Revenues fell for some power sector companies in the study countries because of the global financial crisis. Industrial production plummeted, fueling a drop in demand for electricity (see table 1.1). To protect certain customers, governments postponed tariff increases. In Ukraine, for exam-ple, the government capped tariffs for all customers and moved certain industrial customers into the subsidized tariff category. The combination of lower demand and stagnant tariffs reduced revenues for power sector companies in Armenia, Romania, and Ukraine.

Table 1.1:   Growth in Electricity Consumption, Peak Demand and Exports, 2008–2009

	Consumption		*Peak Demand*
	Total	of which, industrial consumption	
Armenia	-7.4%	-22.2%[a]	-13.5%
Kyrgyz Rep	-0.6%	1.8%[a]	Unknown
Romania	-8.2%	-12.4%	-3.4%
Serbia	-1.9%	-5.8%	-3.2%
Ukraine	-8.7%	-20.2%[a]	-12.4%

Source: Data from utility companies and relevant government agencies.

a. Data available for first two quarters of 2009 only.

Costs rose for many power sector companies, and revenue growth declined (see figure 1.2). Currencies depreciated in all of the study countries, ranging from 15 percent in the Kyrgyz Republic to 36 percent in Ukraine. Depreciation meant that anything that needed to be paid for in foreign currency—fuel imports and foreign currency–denominated debt—cost more. Operating costs increased in three of the study countries—Armenia, the Kyrgyz Republic, and Ukraine—as a result.

Figure 1.2:   Change in Power Sector Revenues, 2007–2009[a]

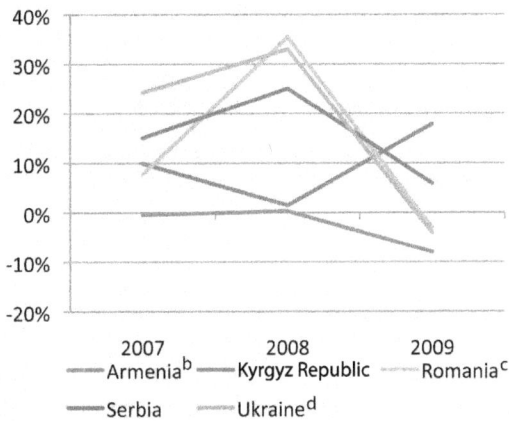

Source: Data from utility companies and relevant government agencies.

a. Calculated as sum of sector companies' revenues in local currency.

b. Armenia: For 2009, shows year-on-year change for first 3Q; no data available for Vorotan or ENA.

c. Romania: State-owned companies only (excluding Hidroelectrica).

d. Ukraine: State-owned thermal power plants only.

Government policy measures—enacted in response to the financial crisis—further affected electricity producers' costs. In Romania, for example, the government allowed gas-fired thermal power plants (TPPs) to purchase discounted gas from Romgaz, the majority state-owned gas company, leading to a decrease in fuel costs at gas-fired TPPs. In Ukraine, the government required that state-owned TPPs buy coal from the state coal mining company at costs higher than available in the market (see box 1.1).

Box 1.1

Impact of the Financial Crisis on Profitability of State-Owned TPPs in Ukraine

In order to support lagging demand for coal during the crisis period, the Cabinet passed resolutions in October 2008, April 2009, and December 2009 requiring state-owned TPPs to purchase coal from SE "Coal of Ukraine" (the state-owned coal mining company). By the end of 2009, a recovery in steel production led to a recovery in the demand for coking coal. Supply began to fall behind demand. Because of the requirement (still in place at the time) that state-owned TPPs buy coal from state-owned mines, prices increased and coal shortages emerged. NAC ECU (the state-owned company responsible for TPPs) had to take on additional short-term loans to pay for increased fuel expenditures.

The combination of increased fuel and financing expenditures led to significant deterioration in the financial performance of state-owned TPPs in the first quarter of 2010. As a result, NAC ECU experienced net losses from February to April of 2010.

Source: NAC ECU.

Falling revenues and rising costs have affected the profitability of power sector companies in many of the study countries. Profit margins declined in almost all segments of the power sector in 2008 and continued to fall in Armenia, Ukraine, and for some companies in Romania in 2009. Some power sector companies experienced negative net income in 2008 and 2009.

Large Investment Needs

Power sector investment needs in the ECA region loomed large before the financial crisis (table 1.2). Large capital expenditure (CAPEX) backlogs existed before the crisis for two primary reasons. First, large amounts of Soviet-era infrastructure needed to be replaced or rehabilitated because of years of under-maintenance or because it had reached the end of the design life. Second, CAPEX plans in many countries were often overstated and not implemented on schedule. Five-year power sector investment needs in the study countries now represent 10 to 40 times the level of investment made between 2007 and 2009.

Table 1.2:   Investment Needs in the Study Countries, 2009–2015, US$ millions

	Investment needs	Secured/ expected financing	Investment gap
Armenia	6,840	984.4	5,855
Kyrgyz Republic	3,573	510.8	3,062.2
Romania	14,665.2	Unknown	Unknown
Serbia	7,722	972-4,381[a]	3,341-6,750[a]
Ukraine	37,655.5	6,825.1	30,830.4

Source: Data from utility companies and relevant government agencies.

a. Depends on whether Serbia can secure strategic partners for construction of new capacity.

Although the overall size of investment needs remains the same as before the financial crisis, the crisis created a window of opportunity for meeting investment targets. The drop in electricity demand delayed by a few years the need for new generation capacity in several of the study countries. Serbia and Ukraine have an additional four- to six- year window, respectively, in which they can make investments in new capacity before an electricity shortage sets in. In Armenia, the financial crisis did not delay the expected supply-demand gap, but did reduce the expected size of the gap.

Limited Available Financing

Securing the financing needed to meet investment targets has become even more difficult in the aftermath of the financial crisis. Financing for power sector projects has become more limited in three important ways. First, the poor financial performance of power sector companies has reduced their ability to fund CAPEX from their own revenues, or secure additional debt or equity financing. Second, the financial crisis has constrained the ability of commercial banks and equity investors to invest in new projects. Capital constraints and higher country and market risks have forced financial institutions to tighten lending requirements and have made foreign investors more risk averse. Third, the financial crisis has limited governments' ability to borrow. The study countries show higher budget deficits and higher public debt, which will limit governments' abilities to finance CAPEX in publicly owned power projects.

The Financial Crisis in Perspective

The impacts of the financial crisis on the financial performance of power sector companies and on the availability of financing should not mask the more endemic problems facing power sectors in the study countries. Investment gaps were large before the crisis and underinvestment common. Similarly, commercial bank financing and private investment were limited before the crisis hit.

Power sector companies' abilities to achieve investment plans and access financing before and, to some extent, during the crisis depended largely on each country's regulatory environment. In three of the study countries (the Kyrgyz Republic, Serbia, and Ukraine) where tariff decisions remained highly politicized, power sector companies had chronic difficulties meeting their investments needs before the crisis. Private sector participation was largely absent from their power sectors, and commercial lending was limited to meeting working capital needs, just as it is now. Investment plans were rarely met. In contrast, in Armenia and Romania sector governance and regulation supported more realistic investment planning.[1]

Stimulating Investments After the Crisis

In the wake of the financial crisis, governments need to focus on funding the most critical projects. This will require governments to do the following:

• Prioritize public spending. With smaller public budgets and scarcer commercial lending, governments will need to prioritize power sector investments carefully. In all of the study countries, energy efficiency is a least cost solution that can postpone the emerging supply-demand gap. Governments will also need to carefully balance capital expenditures—taking into consideration life-cycle costs—with operating and maintenance expenditures as some operating expenditures, particularly fuel costs, continue to grow. This balancing requires a consideration of the tradeoffs between new investment and expenditure on maintenance needed to preserve existing infrastructure. Table 1.3 shows a prioritization of short- and long-term investments in each of the study countries based on criteria of supply reliability, affordability, and compliance with EU regulations.[2]

Table 1.3:   Short- and Long-term Priority Investments in Each Country

	Short-term (1–3 years)	Long-term (4–7 years)
Armenia	Transmission rehabilitation	Construction of new NPP & RE capacity
Kyrgyz Republic	Urgent rehabilitation to improve baseload capacity for upcoming winter	Transmission rehabilitation
Romania	Environmental upgrades of TPPs; re-launching nuclear company; distribution rehabilitation; transmission connections for RE and interconnections	New capacity (conventional thermal, nuclear, hydro, and wind), transmission and distribution rehabilitation
Serbia	Environmental upgrades of TPPs, transmission interconnections, and distribution rehabilitation	New capacity, transmission and distribution rehabilitation
Ukraine	Rehabilitation of HPPs; Rehabilitation of TPPs	Service life extension of NPPs

Source: Authors.

Note: These priority investments are based on the criteria and methodology described in further detail in Appendix B and do not reflect the World Bank's investment strategy in the study countries. NPP = nuclear power plant, RE = renewable energy, TPP = thermal power plant.

- Create a more attractive environment for investment. A legal and regulatory environment that supports a financially viable sector is essential for attracting private investors. It is critical to have and to apply laws and regulations that allow power sector companies to: (i) recover their full costs of service, including the reasonable capital expenditure they planned, and the costs of financing that capital expenditure; (ii) provide predictability in the approval process for investment plans, so that power sector companies are indeed able to plan investment in a rational way; and (iii) recover all of the revenues on electricity sold, by giving them the ability to disconnect nonpaying customers.

The World Bank is well placed to support governments in the case study countries as they look to prioritize public spending and further stimulate private sector participation in the sector. World Bank loans for physical infrastructure may help the government make urgent investments needed for reliability, security, and sustainability of the sector. Advisory service or technical support in implementing legal, regulatory, or institutional changes can attract private sector participation and improve capital expenditure planning in each of the study countries. Additionally, partial risk and partial credit guarantees can help lower the cost of financing and leverage private sector financing that otherwise might not be available.

Notes

1. Recent government actions in Romania, however, have undermined the independence and credibility of the regulator and threaten to undo the achievements of regulatory reform.
2. The prioritization includes only investments that have not yet secured financing and are likely to receive partial or full public (government) funding.

CHAPTER 2

Introduction

Countries in the Europe and Central Asia (ECA) region experienced steadily high electricity demand growth before the global financial crisis. Energy security and supply reliability were a growing concern for policy-makers and planners, as much of the under-maintained, Soviet-legacy power sector infrastructure was in urgent need of replacement. Leading up to the crisis, many countries in the region faced imminent and serious energy supply problems, but had limited funding with which to confront them (see box 2.1).

The global financial crisis hit economies in the ECA region harder than any other region. Gross domestic product (GDP) declined in all of the case study countries in 2009, except in the Kyrgyz Republic, where growth stagnated. The decline in GDP reduced tax revenues, fostering an increase in government budget deficits and public debt. Local currencies also depreciated, most severely in Ukraine and Armenia where they lost one-third and one-sixth of their values against the U.S. dollar, respectively.

The macroeconomic effects of the financial crisis had a variety of follow-on effects in the power sector. Electricity demand declined in all of the case study countries with the decline in economic output. On one hand, this worsened the financial performance of power sector companies, reducing their ability to attract financing as well as their ability to generate cash for investment. On the other hand, the drop in demand temporarily delayed some of the need for new investment.

This report builds on earlier World Bank work in the region and the sector by focusing on what has happened in ECA countries' power sectors as a result of the financial crisis. It identifies the impacts of the financial crisis on power sectors in the region by focusing on five countries (the case study countries): Armenia, the Kyrgyz Republic, Romania, Serbia, and Ukraine.

Box 2.1

World Bank "Lights Out?" Report Highlights Energy Outlook in ECA

In March 2010, the World Bank released its energy flagship report for Eastern Europe and Central Asia (ECA) titled *Lights Out? The Outlook for Energy in Eastern Europe and Central Asia*. Key findings from this report related to the power sector are as follows:

- Threat of energy shortages. The ECA region could face energy shortages in the next five to six years if needed investments are not made.
- Energy trends reflect economic trends. Production and consumption of energy historically reflect economic performance in the ECA region. The global financial crisis of 2008 accordingly dampened energy demand, creating temporary breathing room before energy supply again becomes a major concern.
- Large investment needs. To stave off electricity shortages, the region needs more than US$ 1.5 trillion investment in power sector infrastructure in the next 20 to 25 years.
- Need to attract private financing. The level of investment required in the energy sector cannot be financed by the public sector alone. However, attracting private sector financing will require changing the investment climate.
- Take action now. With large investment needs and long lead times to implement energy projects, governments need to take action now to attract investment.
- Energy efficiency is least-cost investment. Each additional US$ 1 invested in energy efficiency can avoid more than $2 in production investment. Government plays a major role in removing barriers to investment in energy efficiency.

Source: World Bank. 2010. *Lights Out? The Outlook for Energy in Eastern Europe and the Former Soviet Union.* Washington, DC: World Bank.

The report identifies the impacts of the financial crisis on the study countries' power sectors in order to:

- Identify actions governments can take to prioritize public spending in the sector in the short term (up to 3 years) and long term (4–7 years),
- Identify options and government actions required to leverage private investment in the sector, and
- Recommend ways in which the World Bank can support governments in their actions.

Figure 2.1 illustrates the report's approach to these objectives.

Figure 2.1:   Objectives and Approach of the Report

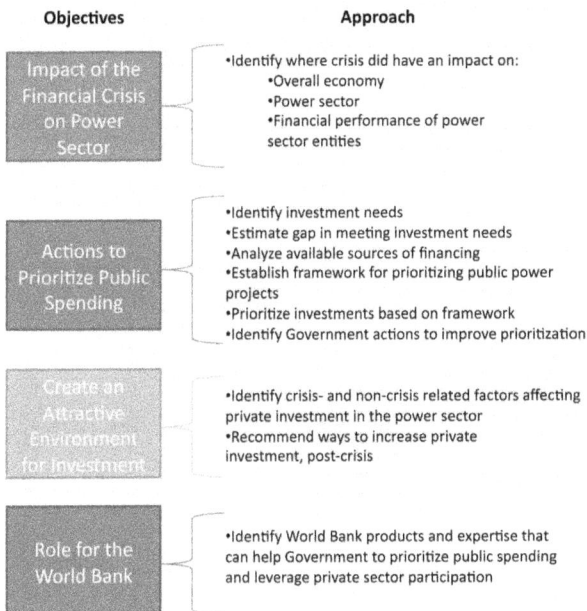

Objectives	Approach
Impact of the Financial Crisis on Power Sector	•Identify where crisis did have an impact on: •Overall economy •Power sector •Financial performance of power sector entities
Actions to Prioritize Public Spending	•Identify investment needs •Estimate gap in meeting investment needs •Analyze available sources of financing •Establish framework for prioritizing public power projects •Prioritize investments based on framework •Identify Government actions to improve prioritization
Create an Attractive Environment for Investment	•Identify crisis- and non-crisis related factors affecting private investment in the power sector •Recommend ways to increase private investment, post-crisis
Role for the World Bank	•Identify World Bank products and expertise that can help Government to prioritize public spending and leverage private sector participation

Source: Authors.

The report is structured as follows:

- **Chapter 3** describes the impacts of the financial crisis on the economies of the case study countries, and on their power sectors specifically.

- **Chapter 4** estimates the investment gap in each of the case study countries' power sectors and identifies how the financial crisis affected their abilities to close the gap.
- **Chapter 5** summarizes our conclusions on the impacts of the financial crisis on the case study countries' power sectors.
- **Chapter 6** recommends what the case study countries' policymakers can do to cope with the impacts of the financial crisis. The section includes recommendations for prioritizing public spending with limited funds, and creating a more attractive environment for private investment in the power sector. The section also identifies a possible role for the World Bank in supporting governments in implementing the section's recommendations.

The information in the appendixes supports the analysis of each country's power sector and the prioritization of new power sector investments.

CHAPTER 3

Impacts of the Financial Crisis

The macroeconomic effects of the financial crisis had a direct impact on the power sectors of the case study countries and the financial performance of power sector companies. Figure 3.1 summarizes the effects of the financial crisis on each case study country.

Figure 3.1: Impacts of the Financial Crisis in Each of the Case Study Countries

		Armenia	Kyrgyz Republic	Romania	Serbia	Ukraine
Macro effects	GDP	●	●	●	●	●
	Budget Deficit	●	●	●	●	●
	Currency	●	●	●	●	●
Power sector effects	Demand	●	○	●	●	●
	Tariffs	○	○	●	●	●
Effects on financial performance of power sector entities	Revenues	●	○	●	○	●
	OPEX	○	○	●	○	●
	Net Income	◐	○	○	○	○
	Debt Service	○	○	○	○	●
	CAPEX	○	○	○	○	○

● = Effect of Financial Crisis   ○ = Effect of Financial Crisis and Other Factors   ○ = Not Effect of Financial Crisis

Source: Authors.

This section explains the results shown in figure 3.1 by analyzing the macro-economic impacts of the crisis and how these impacts flowed through to the power sector and financial health of power sector companies.

Macroeconomic Effects

The macroeconomic effects of the financial crisis affected the power sectors in Armenia, the Kyrgyz Republic, Romania, Serbia, and Ukraine in three ways:

- Gross domestic product slowed or declined, leading to a decrease in demand for electricity.
- Currencies depreciated, leading to higher costs for imported goods, including equipment, materials, and fuel.
- State budget deficits increased, public debt levels increased, and debt ratings deteriorated, tightening the fiscal space available for capital expenditure (CAPEX).

Gross domestic product slowed or declined

GDP declined in all of the case study countries in 2009 except in the Kyrgyz Republic, where growth slowed from 7.9 percent to 0.9 percent. The crisis hit export-oriented and energy-intensive sectors the hardest in all of the case study countries. Industrial production declined 19 percent

Figure 3.2:   Percent Change in GDP, 2006–2010 (projected)

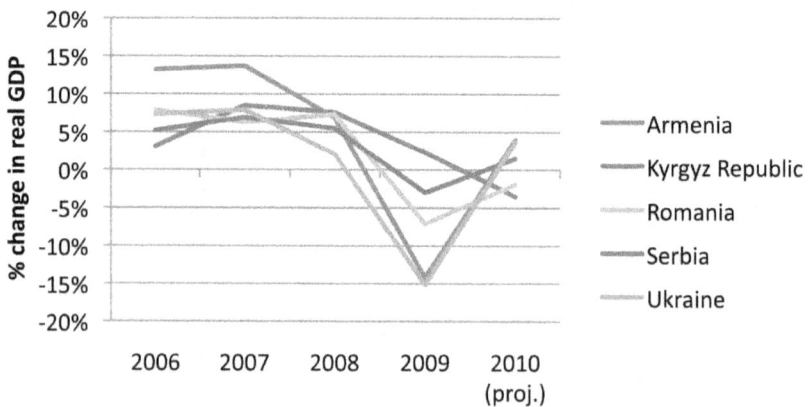

Source: International Monetary Fund. October 2010. *World Economic Outlook: Recovery, Risk, and Rebalancing.* Washington, DC: IMF.

in the Kyrgyz Republic (first half of 2009) and 33 percent in Ukraine (first three-quarters of 2009). In Armenia and Serbia, construction declined 52 percent and 17 percent, respectively. Sectors hardest hit in Romania included mining, which declined 54 percent, and metallurgy, which declined 44 percent in 2009.

Growth is expected to recover moderately (1.3–3.7 percent) in 2010 and improve further in subsequent years. Figure 3.2 depicts the decline in GDP growth in 2009 and how growth is expected to rebound slightly in 2010.

Currencies depreciated

Local currencies depreciated in all of the case study countries. The impact was most severe in Ukraine, where the hryvnia lost more than one-third of its value in the fourth quarter of 2008. In Armenia, the Central Bank let the dram depreciate by 16 percent against the U.S. dollar in March 2009. Table 3.1 shows the average exchange rates for each of the five case study countries in 2008 and 2009 and their depreciation against the dollar over that period.

Table 3.1:   Exchange Rates in Case Study Countries, 2008–2009

	Local currency	2008	2009	Depreciation (against US$)
Armenia	Dram (AMD)	306	363	16%
Kyrgyz Republic	Soum (KGS)	36.6	43	15%
Romania	Lei (RON)	2.5	3.1	19%
Serbia	Dinar (RSD)	55.7	67.5	17%
Ukraine	Hryvnia (UAH)	5.05	7.95	36%

Source: IMF Country Reports.[1]

Budget deficits and public debt increased

Budget deficits and public debt levels increased in 2009 in all case study countries because of the decline in GDP and resulting reduction in tax revenues. Ratings agencies consequently downgraded all of the case study countries with rated sovereign debt (Armenia, Romania, Serbia, and Ukraine). Table 3.2 shows how budget deficits, levels of public debt, and debt ratings changed in the five case study countries from 2008 to 2009.

Table 3.2:   Tax Revenues, Budget Deficit, and Public Debt (% of GDP), 2008–2009

	State budget deficit		Public debt		Debt rating
	2008	2009	2008	2009	
Armenia	1.2	8.0	16.2	40.6	In August 2009, Fitch downgraded the long-term foreign and local currency Issuer Default Ratings (IDR) for Armenia from 'BB' to 'BB-' and downgraded the Country Ceiling from 'BB+' to 'BB.'
Kyrgyz Republic	0	3.7	48.5	59.4	Debt not rated
Romania	4.8	7.4	19.5	28.2	In October 2008, Fitch downgraded Romania's long-term foreign currency debt from 'BBB' to 'BB+'; the rating has since been maintained.
Serbia	2.6	4.1	33.4	35.6	Standard & Poor's has maintained Serbia's sovereign debt rating of 'BB-' since 2007, although outlook shifted from "positive" in 2007 to "negative" in March 2008 and returned to "stable" in December 2009.
Ukraine	3.2	6.2	19.9	34.6	In February 2009, S&P cut Ukraine's long-term foreign currency rating two levels to 'CCC+'.

Source: IMF Country Reports.

Effects on Power Sector Financial Performance

The macroeconomic effects of the financial crisis had a direct impact on the power sectors of most of the case study countries. Specific effects in the power sector included the following:

- A decrease in electricity consumption resulting from the decline in GDP.
- A delay in the supply-demand gap resulting from the decrease in electricity consumption.
- Delays in plans to hike tariffs for certain customer groups.
- Declining revenues in Armenia, Romania, and Ukraine because of lower demand and postponed tariff hikes.

- Higher operating expenditures in Armenia, the Kyrgyz Republic, and Ukraine. Rising fuel costs, amplified by the currency depreciation, led to increased operating expenditures in most of the case study countries.
- Higher debt service costs for some companies because of currency depreciation.
- Declining profitability in all countries because of declining revenues and rising costs.

The following subsections look at each of these impacts in further detail.

Lower electricity demand

Electricity consumption decreased in all of the case study countries in 2009, ranging from 0.6 percent in the the Kyrgyz Republic to 8.7 percent in Ukraine. Peak demand dropped in four of the five case study countries, ranging from 3.2 percent in Serbia to 13.5 percent in Armenia (see figure 3.3).[2]

The decrease in industrial output drove much of the decrease in electricity consumption in Armenia, Romania, Serbia, and Ukraine. For example, industrial consumption in Armenia, which accounts for roughly 25 percent of the country's total electricity consumption, dropped 22 percent in the first two quarters of 2009. Industrial consumption in Ukraine, which accounts for more than 50 percent of the country's total electricity consumption, dropped 20.2 percent in the first two quarters of 2009.

Figure 3.3:   Quarterly Change in Electricity Consumption, 2007–2009

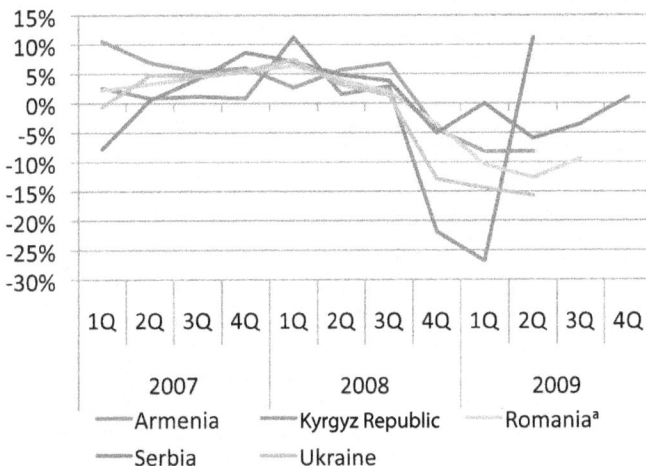

Source: Data from utility companies and relevant government agencies.

a. Change in 2007 consumption in Romania based on annual data.

In the Kyrgyz Republic, it is hard to differentiate the impacts of the financial crisis on electricity demand from the impacts of a concurrent energy crisis. Box 3.1 briefly describes the energy crisis in the Kyrgyz Republic and explains why it is difficult to differentiate these impacts from those of the financial crisis.

Box 3.1

How did the energy crisis affect the power sector in the Kyrgyz Republic?

Low water levels at Toktogul reservoir and an unusually cold winter forced power cuts during the winter months in 2007–08 and 2008–09. The impacts of the power cuts make it difficult to identify the impacts of the financial crisis on three important indicators:

- GDP. Many sectors in the Kyrgyz Republic were negatively affected by the power cuts. One study suggested that a 1 percent decrease in industrial electricity consumption could be associated with a 2.5 percent decrease in GDP and budgetary revenues.[3] It is therefore difficult to determine how much of the reduction in the Kyrgyz Republic's GDP was caused by the financial crisis, and how much by the energy crisis.
- Electricity consumption. The reduced electricity consumption observed during 2008 and 2009 partly resulted from forced power cuts during that period, making it difficult to determine how much of the observed decline in consumption resulted from reduced demand.
- Financial performance of power sector companies. Utilization of the Bishkek and Osh combined heat and power plants, which cost 25 times more to operate than the country's hydro plants, increased 26 percent from 2007 to 2008 to compensate for reduced generation at Toktogul hydropower plant. This contributed to an overall decline in net revenues for the state-owned generation company.

Source: Data from utility companies and relevant government agencies.

Delayed tariff hikes

Plans to increase tariffs were postponed in all of the case study countries during the crisis period, in order to protect certain customer groups. More specifically:

- In Armenia, to neutralize the impact of higher gas prices on retail tariffs, the government waived a portion of the tariff that is meant to provide a return on assets for state-owned power plants.
- In the Kyrgyz Republic, a policy was adopted in 2009 that led to a two fold tariff increase on January 1, 2010. The tariff increase was later reversed by the interim government. Box 4.2 (see chapter 4) describes how the political uprising in April 2010, and subsequent riots in June 2010, affected the power sector in the Kyrgyz Republic.
- In Romania, the regulator maintained tariffs for captive residential customers[4] at 2008 levels and does not plan to increase them until January 2011.
- In Serbia, in 2009 the government postponed any increase in end-user tariffs until March 1, 2010.
- In Ukraine, the government issued a presidential decree in November 2008 setting a moratorium on price increases for natural monopolies, which included distribution companies. As a result, tariffs were capped for all customer groups. The government also moved the mining, metallurgical, and chemical industries into the subsidized electricity tariff category.

Lower revenues

Lower demand and stagnant tariffs led to lower power sector revenues in Armenia, Romania, and Ukraine. The average drop in revenue for the sector ranged from 3 percent in Romania to 8 percent in Armenia. More specifically:

- In Armenia, power sector companies experienced drops in revenue that ranged from 1 to 17 percent during the first three quarters of 2009.
- In Romania, revenues tended to reflect sales. In generation, for example, revenues decreased 12.7 percent at Turceni (a state-owned thermal power plant) in line with a 22.6 percent decrease in generation, compared with Nuclearelectrica (state-owned nuclear generating company), where revenues increased 31 percent in line with a 4.8 percent increase in generation. Revenues were also lower because electricity market prices dropped. Prices on the day-ahead market dropped 22.8 percent in RON (33 percent in EUR).

Sales were also lower in the transmission and distribution segments. Revenues decreased 4 percent at Electrica (the state-owned distribution company), and 17.7 percent at Transelectrica (the transmission service operator), because of a 41.2 percent decrease in balancing market transactions. The price-cap methodology used in the distribution sector means that some revenues covering fixed costs will be recouped in the next tariff revision, but the sector regulator (ANRE) has indicated that it will likely postpone a full revenue "true up" for distribution companies.

- In Ukraine, revenue changes reflected changes in generation and tariff levels. For example, a 21 percent decrease in generation and 1 percent decrease in average tariff levels affected revenues at thermal power plants (TPPs), which on aggregate decreased 4 percent in 2009. Generation decreased for TPPs more than for any other type of generation because the drop in demand shifted the generation mix toward cheaper sources of generation, such as nuclear and hydro, and away from more expensive TPPs.

In the other two case study countries, increased exports (Serbia) and factors unrelated to the financial crisis (the Kyrgyz Republic) led to an increase in sector revenues:

- In Serbia, despite the 3.5 percent drop in consumption, revenues increased 7 percent at EPS (the state-owned generation and distribution company) in 2009. The increase was driven by a 75 percent increase in electricity exports. Electricity exports increased because of:
  - Good hydrological conditions that allowed for increased generation at HPPs,
  - Lower domestic consumption, which increased electricity available for export, and
  - The currency depreciation, which made the cost of electricity from Serbia relatively cheaper than in neighboring countries.

In contrast, revenues at EMS (the transmission system operator) dropped 6 percent, in line with the decrease in domestic consumption, since EMS does not benefit from increased export sales volumes.

- In the Kyrgyz Republic, revenues increased 18 percent in 2009 as generation recovered from the winter 2007/08 power cuts. Other factors, including lower commercial losses, also contributed to revenue increases in the Kyrgyz Republic. Commercial losses decreased by 28 percent in 2008 and by 12 percent in 2009.[5]

Figure 3.4 shows how revenues changed from 2007 to 2009 in each of the case study countries.

Figure 3.4:   Change in Power Sector Revenues, 2007–2009[a]

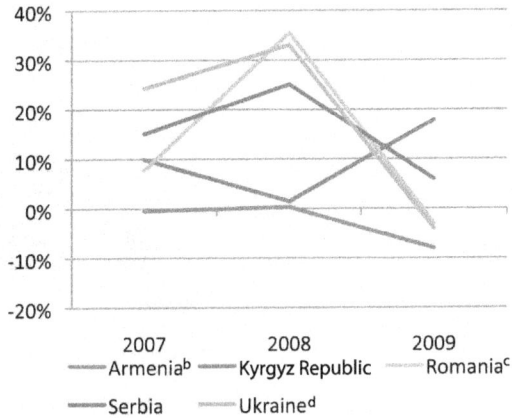

Source: Data from utility companies and relevant government agencies.

a. Calculated as sum of sector companies' revenues in local currency.

b. Armenia: For 2009, shows year-on-year change for first 3Q; no data available for Vorotan or ENA.

c. Romania: State-owned companies only (excluding Hidroelectrica).

d. Ukraine: State-owned TPPs only.

Higher operating costs

The impact of the financial crisis on power sector operating expenditures varied across the case study countries (see figure 3.5). In Armenia, the Kyrgyz Republic, and Ukraine, higher fuel costs led to operating cost increases that ranged from 2 to 16 percent. Fuel costs increased during the crisis period for three reasons:

- Fuel prices increased:
  - In Armenia, natural gas prices increased 40 percent in 2009 and 17 percent in 2010, and the cost of nuclear fuel increased by 35 percent in 2009.
  - In the Kyrgyz Republic, coal prices increased 13 percent and gas prices increased 66 percent in 2009.
  - In Ukraine, coal prices (purchased in local currency) increased 27 percent and gas prices (purchased in foreign currency) increased 22 percent in 2009.
- National currencies depreciated, which further increased the cost of fuel purchased in foreign currency. Armenia and the Kyrgyz Republic purchase all of their fuel in foreign currency, and Ukraine purchases all of its natural gas in foreign currency.

Figure 3.5:   Change in Operating Expenditures, 2007–2009[a]

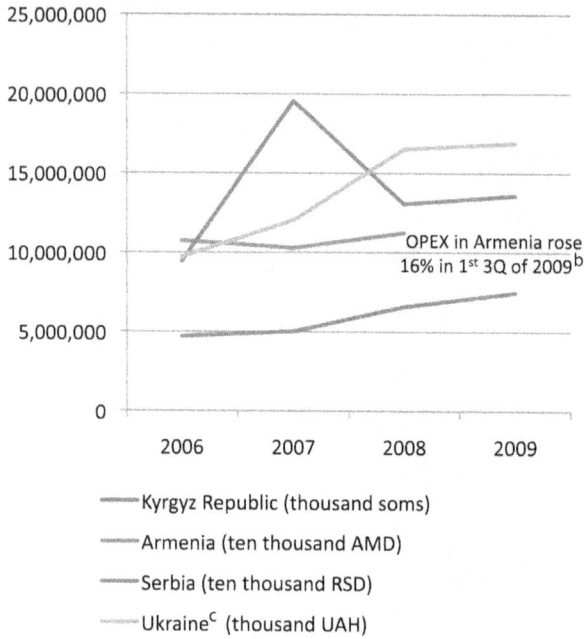

Source: Data from utility companies and relevant government agencies.

a. Calculated as sum of sector companies operating expenditures; No time series data available for Romania.

b. No data available for Vorotan or ENA in 2009.

c. State-owned TPPs only.

- Thermal generating companies increased fuel purchases despite decreasing consumption:
  - In the Kyrgyz Republic, the energy crisis caused by low water levels at the Toktogul reservoir forced increased generation and, hence, increased fuel consumption, at Bishkek CHP.
  - In Ukraine, in an effort to support state-owned coal mines during the crisis, the government required that state-owned TPPs buy excess coal from the state coal mining company even as generation at TPPs declined.

Meanwhile, in Romania and Serbia operating expenditures decreased in 2009:

- In Romania, companies decreased nonfuel operating expenditures (5–20% at TPPs) and cut employment in an effort to balance their budgets. Additionally, in response to the crisis, the government allowed gas-fired TPPs to purchase discounted gas from Romgaz, the majority state-owned gas company, leading to a decrease in fuel costs at gas-fired TPPs.

- In Serbia, operating expenditures increased 5 percent at EPS (the state-owned generation and distribution company) as production increased to serve the export market. In contrast, operating expenditures decreased 10 percent at EMS (the transmission company) as domestic consumption decreased.

Weaker ability to service debt

Debt service coverage ratios (DSCR) deteriorated in most of the case study countries during the crisis period:[6]

- In Armenia, the DSCR at Vorotan HPP (state-owned hydro company) deteriorated in 2008.
- In the Kyrgyz Republic, the DSCR at JSC NESK (state-owned transmission company) has been below 1 since 2006. At JSC ES (state-owned generation company), the DSCR has been below 1 since 2008.[7]
- In Romania, Hidroelectrica (state-owned hydropower company) is not currently meeting World Bank debt covenants. Transelectrica (majority state-owned transmission company) is not currently meeting European Investment Bank (EIB), World Bank, and European Bank for Reconstruction and Development (EBRD) financial covenants related to pretax working ratio and current ratio in 2009, but the covenant breach is not substantial.
- In Serbia, the DSCR at EMS (state-owned transmission company) fell from 2 to 0.85 in 2009 even though debt service costs decreased 18.5 percent.
- In Ukraine, the DSCR was below 1 at most state-owned TPPs in 2009.

Debt service coverage ratios deteriorated because companies took on more debt (short- and long-term), currency depreciations increased the cost of servicing debt, or net income dropped. More specifically:

- In Armenia, short-term debt for the distribution company increased 5.4-fold in 2008 and 24-fold for Sevan-Hrazdan HPP.
- In the Kyrgyz Republic, debt service (as a percentage of total costs) increased from 6 percent to 25 percent for JSC ES (the generation company) as a result of financing Kambarata-2 HPP.
- In Ukraine, short-term debt at TPPs (except Zakhidenergo) increased over 50 percent between January 1, 2008, and January 1, 2009, in line with currency depreciation in 4Q 2008. Additionally, Ukrhydrenergo

(state-owned hydropower plant) had to secure additional US$ 60 million financing from the World Bank because after the currency depreciation it could no longer finance the U.S. dollar portion of an existing World Bank loan.[8]

The following section describes in more detail how net income changed during the crisis period for all segments of the power sector in each of the case study countries.

Lower net income

Declining operating margins and profit margins, and negative net income emerged in 2008 and continued into 2009 at many companies in each of the case study countries. In some cases, these trends can be considered an impact of the financial crisis: revenues decreased as a result of the drop in demand and costs increased as a result of the currency depreciation.

In other cases, however, changes in net income can be attributed to other causes unrelated to the financial crisis. For example, Serbia's EPS (the state-owned generation and distribution company) has experienced net losses in recent years because of an asset revaluation in 2007. Table 3.3 shows how operating margins and profit margins changed from 2006 to 2009 in each segment of the case study countries' power sectors.

Table 3.3:   Operating and Profit Margins, 2006–2009

Country	Segment		2006	2007	2008	2009
	Operating Margins					
Armenia	Generation		2%	2%	0%	no data
	Transmission		4%	4%	1%	
	Distribution		1.6%	2.2%	2.3%	
Kyrgyz Republic	Generation		15%	17%	-21%	-10%
	Transmission		21%	10%	-1%	12%
	Distribution		66%	65%	61%	61%
Ukraine	Generation (state-owned TPPs only)		7%	7%	4%	-2%
	Profit Margins					
Romania	Generation	Turceni	4.1%	7.6%	1.6%	1.4%
		Rovinari	5.3%	12%	2.1%	2.6%
		Craiova	4%	1%	0.3%	0.2%
		Termoelectrica	-51.5%	104.7%	-53.9%	-39%
		Nuclearelectrica	41.9%	41.0%	5.2%	4.5%
		Hidroelectrica	no data	no data	3.3%	2.6%
	Transmission		11.7%	2.6%	1.7%	0.7%
	Distribution		-149.7%	116.6%	45.4%	8.3%
Serbia	EPS (G,D)		17%	-89%	-19%	-6%
	EMS (T)		20%	9%	2%	4%

Source: Data from utility companies and relevant government agencies and for Romania, www.doingbusiness.ro, utility company profiles retrieved August 16, 2010, http://mcir.doingbusiness.ro.

Note: Data only available to calculate operating margins in Armenia, Kyrgyz Republic, and Ukraine and profit margins in Romania and Serbia.

Notes

1. Armenia: IMF. July 2010. Country Report No. 10/223; the Kyrgyz Republic: IMF. October 2010. Country Report No. 10/336; Romania: IMF. September 2010. Country Report No. 10/301; Serbia: IMF. October 2010. Country Report No. 10/308; Ukraine: IMF. August 2010. Country Report No. 10/262.
2. No data available on the change in peak demand in the Kyrgyz Republic.
3. Addyshev, Nurlan. "Industrialists say that power cuts affect production volumes and GDP." Business AKIpress, September 3, 2008.
4. A captive customer is defined as: "An electricity customers, who for technical, economic or regulatory reasons, is unable to purchase electricity from the supplier of his choice," from the Liberalization of the Electricity Market in Romania – Glossary of terms. National Energy Regulatory Agency of Romania (ANRE), http://www.anre.ro/informatii.php?id=741
5. Commercial losses calculated as percentage of total generation.

6. Debt service is a subcategory of operating expenditures. We treat it separately in this paper given: (i) the observed impact of the financial crisis on debt service in some of the case study countries and (ii) the impact that ability to meet debt covenants has on future availability of financing for investments.

7. A debt service coverage ratio below 1 indicates that a company lacks sufficient income from operating activity to cover all debt payment obligations. If net income declines or if the cost of servicing debt increases, the debt service coverage ratio deteriorates.

8. The World Bank provided a loan to Ukrhydrenergo (UHE) of US$ 106 million in 2005. The loan agreement required UHE to co-finance US$ 268 million of the project, of which US$ 18 million had to be financed in foreign currency. UHE struggled to finance the foreign currency component of project costs after the 2008 currency depreciation. In response, the World Bank provided an additional US$ 60 million in financing in May 2009.

CHAPTER 4

Financing Needs

Power sectors in the case study countries had large investment needs (an investment gap) before the global financial crisis, and a scarcity of funds to meet those needs. The financial crisis has weakened the financial condition of public and private companies, making them less creditworthy and less able to fund investment from cash generated internally. The crisis has therefore made it more difficult to fill the investment gap. This chapter quantifies the investment gap facing the power sector in each of the case study countries and then analyzes sources of financing available in the postcrisis period.

Investment Gap

Investment gaps existed before the financial crisis in most of the case study countries. Large amounts of Soviet-era infrastructure must be replaced or rehabilitated within the next five to ten years because of years of under-maintenance or because they have reached the end of their design life.[1] Most of the case study countries had large capital expenditure backlogs before the financial crisis and continue to have them.

Power sector companies in the Kyrgyz Republic, Serbia, and Ukraine have a history of missing their CAPEX targets. Power sector companies in Armenia and Romania, in contrast, regularly meet their CAPEX targets. Figure 4.1 through figure 4.5 show how CAPEX plans for generation,

Figure 4.1:   Actual and Planned CAPEX in Armenia, 2006–2011

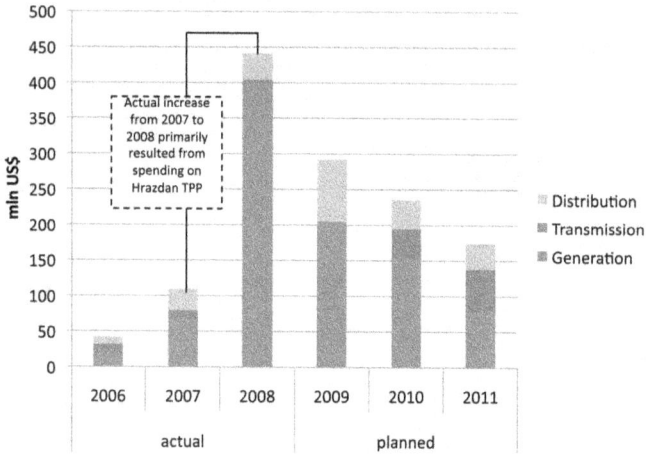

Source: PSRC, Ministry of Energy and Natural Resources of RA.

Figure 4.2:   Actual and Planned CAPEX in the Kyrgyz Republic, 2006–2012

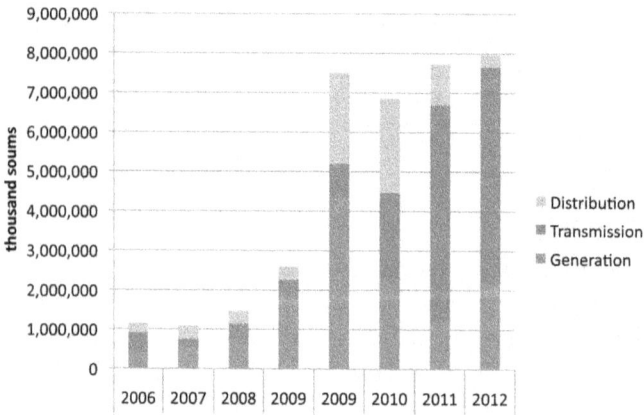

Source: Data provided by National Regulator.

transmission, and distribution compare to actual CAPEX in recent years for each of the case study countries. The figures also show CAPEX plans for future years.

Figure 4.3:   Actual and Planned CAPEX in Romania, 2006–2011[a]

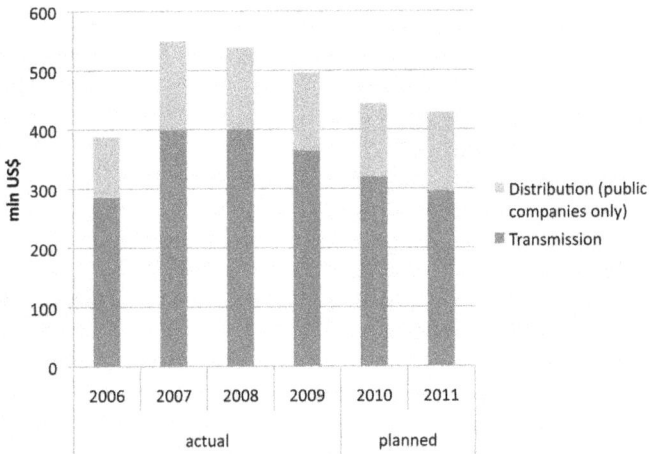

Source: Investment plans of Transelectrica (majority state-owned transmission company), Electrica Muntenia Nord, Electrica Transilvania Nord, and Electrica Transilvania Sud (state-owned distribution companies).

a. No data available for generation.

Figure 4.4:   Actual and Planned CAPEX in Serbia, 2006–2012

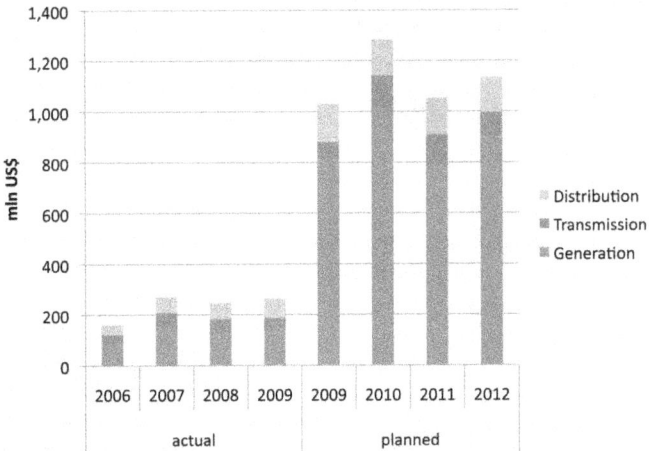

Source: EPS and EMS.

Figure 4.5:   Actual and Planned CAPEX in Ukraine, 2006–2011

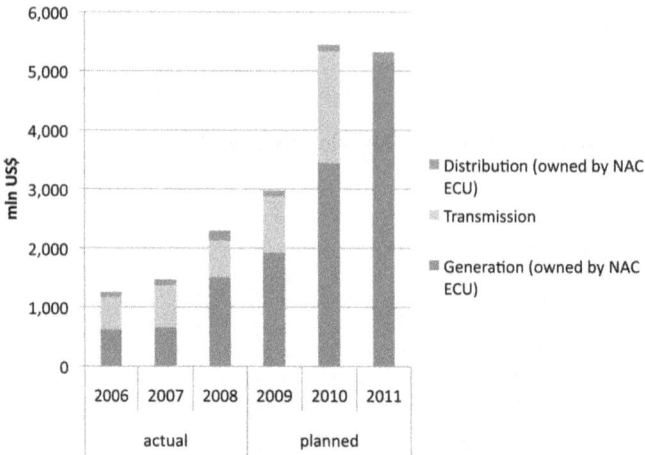

Source: NAC ECU and Ukrenergo.

A significant share of the CAPEX required in Romania, Serbia, and Ukraine is for investment in environmental upgrades and renewable energy needed to comply with European Union (EU) regulations (see box 4.1).

The financial crisis had little impact on the overall size of investment needs or the size of the investment gap, but it did postpone the need for some new generating capacity. The drop in electricity demand in 2009 has delayed—by a few years—the need for new generating capacity in several of the case study countries.[2]

Figure 4.6 through figure 4.10 show the emerging supply-demand gaps in each of the case study countries.

In Armenia, the investment gap is forecasted for 2017, but the decrease in demand reduced the size of the gap in meeting peak demand from roughly 1100 MW to 518 MW to 918 MW, depending on assumptions about demand growth.

In the Kyrgyz Republic, generation and consumption dropped, but they are expected to return to historic average levels by 2012.[3]

In Romania, the gap in meeting peak demand and reserve margin emerges if no new capacity is built by 2017. This gap in meeting peak demand is much larger if old TPPs are not upgraded. If hard coal, lignite, gas, and oil TPPs are shut down because they do not comply with EU directives, the gap in meeting peak demand and reserve by 2017 will be 9,010 MW and 12,777 MW, respectively.

Box 4.1

How do EU Directives affect investments in Romania, Serbia, and Ukraine?

European Union Directives require investments in environmental upgrades of TPPs in Romania, Serbia, and Ukraine and new renewable energy capacity in Romania. EU Directive 2001/80/EC on large combustion plants (LCPs) imposes emission reduction requirements on existing large power plants. EU Directive 2009/28/EC requires investment in renewable energy. These directives affect power sector investments in the case study countries as follows:

- As a member of the EU, Romania must invest in environmental upgrades for 52 percent of its installed capacity by 2013 and invest heavily in renewable energy capacity to meet the country's EU target to supply 24 percent of energy consumption from renewable energy by 2020.

- As a member of the Energy Community of South East Europe, Serbia has a legal obligation to comply with the LCP directive. This requires environmental upgrades of 3,409 MW of TPPs in Serbia.

- Ukraine's parliament ratified the Energy Community Treaty on December 15, 2010, making thermal power plants legally obligated to comply with the LCP directive.

Figure 4.6:   Peak Demand and Available Capacity in Armenia, 2006–2019

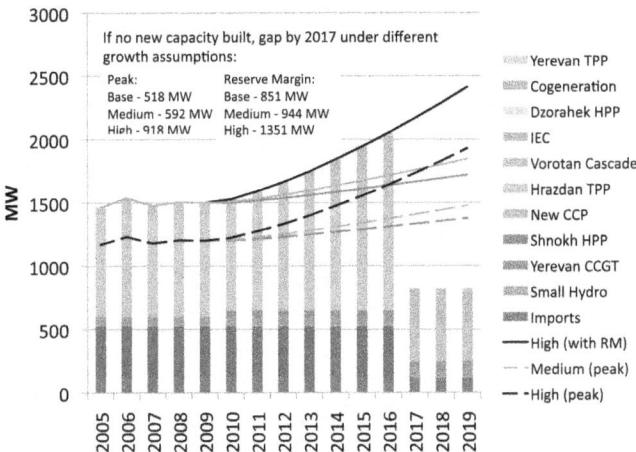

Source: Demand forecast based on World Bank Armenia Energy Issues Note.

Note: Annual demand growth assumptions: Base scenario = 1.53%; Medium scenario = 2.28%; High scenario= 5.27%; RM = reserve margin.

Figure 4.7:   Generation and Consumption in the Kyrgyz Republic, 2006–2020

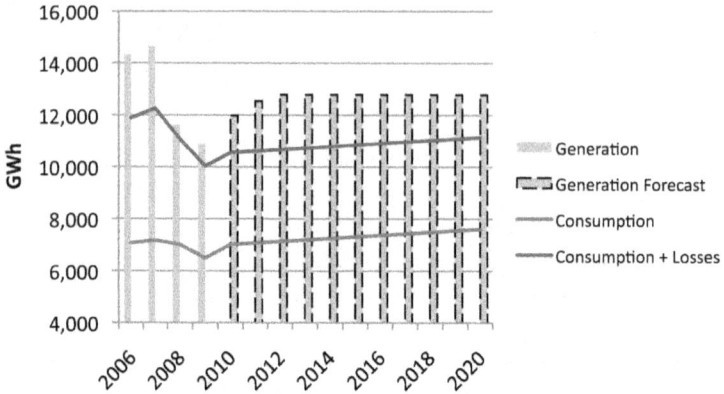

Source: Consumption: Assumes historic average annual growth of 1%; Generation Forecast: State Department on Regulation of the Fuel and Energy Sector.

Note: No data were available on peak demand and available capacity for the Kyrgyz Republic. Generation and consumption forecast does not show a gap in meeting consumption in the Kyrgyz Republic. However, lack of available hydro capacity in winter creates a seasonal gap in meeting consumption and demand not demonstrated in figure.

Figure 4.8:   Peak Demand and Available Capacity in Romania, 2007–2017

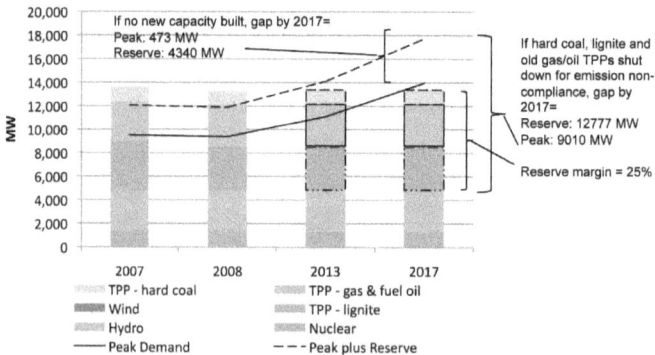

Source: Data from utility companies and relevant government agencies and Transelectrica for demand forecast.

In Serbia, the drop in electricity demand is expected to postpone the need for new winter peaking capacity by as much as six years (from 2013 to as late as 2019, depending on assumptions about demand growth).

Figure 4.9:   Peak Demand and Available Capacity in Serbia, 2007–2025

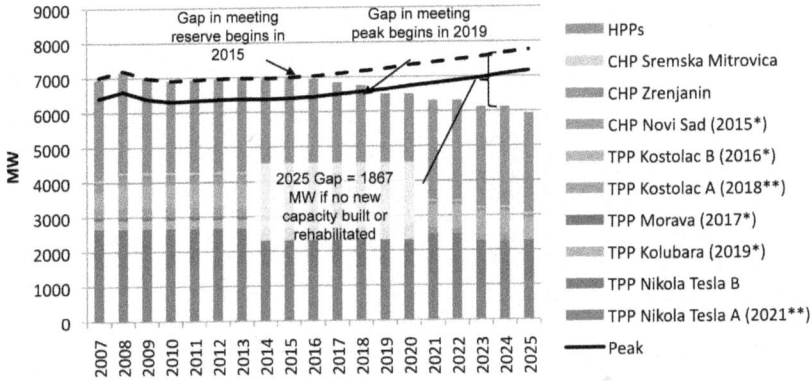

Source: EPS.

In Ukraine, the drop in electricity demand delayed the emergence of a supply gap by as much as four years (from 2015 to as late as 2019, depending on assumptions about demand growth).

Figure 4.10:   Peak Demand and Available Capacity in Ukraine, 2008–2029[a]

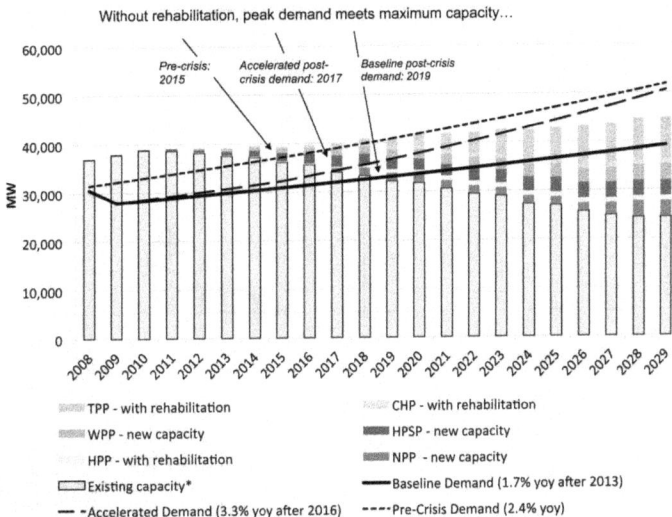

Source: IMEPower calculation based on precrisis rehabilitation schedule.

a. Assumes continuation of existing capacity beyond 2010 except for TPPs. However, continuation of existing capacity will require rehabilitation to prevent drop in available capacity (for CHPs and HPPs) and service life extension for NPPs.

Table 4.1:   Size of Investment Needs and the Investment Gap in the Case Study Countries

	Source and years	Investment needs	Secured/ expected financing	Investment gap	Financing still needed for... [a]
Armenia	Companies' investment plans, 2009-2013; Government energy sector development strategy	6,840	984.4	5,855	• Hrazdan TPP Unit 5: US$ 60 mln (but close to securing the financing) • Sevan-Hrazdan HPP: US$ 40 mln • Replacement of ANPP: US$ 5.5 bln • Lori-Berd and Shnokh HPPs: US$ 250 mln
Kyrgyz Republic	Short-term Energy Sector Development Strategy for 2009-2012	3,573	510.8	3,062.2	• Datka-Kemin 500 kV line and substations: US$ 336 mln • Distribution rehabilitation and metering: US$ 150 mln • Bishkek CHP or Karakeche TPP: US$ 350 mln or 1.2 bln • Kambarata-1: US$ 1.7 bln
Romania	Planned CAPEX for distribution companies, 2009-2011; Reports of private investment plans	14,665.2	Unknown	Unknown	• Environmental upgrade of TPPs: US$ 1,432.2 mln • New wind power plants: US$ 4,728.4 mln • New conventional thermal power plants: US$ 3,654.3 mln • Ongoing rehabilitation of distribution: US$ 1,911.9 mln
Serbia	Investment Plans of EPS & EMS, 2009-2015	7,722	972-4,381[b]	3,341-6,750[b]	• Environmental upgrade of TPPs: US$ 1,039 mln • Construction of new capacity (Kolubara B, TPP Nikola Tesla B3, CHP Novi Sad): US$ 6,428 • Distribution: US$ 1,058 mln
Ukraine	Companies' investment plans, 2009-2011; MFE Action Plans for each segment until 2015	37655.5	6825.1	30830.4	• TPPs: US$ 6,576.6 mln • Nuclear: US$ 5,048.5 mln • CHPs: US$ 2,156.4 mln • Wind: US$ 9,603.5 mln • Transmission: US$ 2,551 mln • Distribution: US$ 4,894.4 mln

a. Includes only the largest investments that still need financing.

b. Depends on whether Serbia can secure strategic partners for construction of new capacity.

Table 4.1 provides an overview of the size of investment needs and the investment gap in each of the case study countries, and highlights some of the major investments needed in the sector.

Investment needs are large in each of the case study countries. Table 4.2 provides a comparison of the size of the investment gap in each country relative to the size of the sector, the state budget, and the overall economy.

Table 4.2:   Comparison of Investment Gap to GDP, State Budget, Sector Revenues, and Sector Capital Expenditures, US$ millions

		Investment gap		GDP	State budget	Gross sector revenues	Total CAPEX
		2010-2015	Annual Average	2008	2008	2008	Annual Average, 2006-2008
Armenia		5,855	976	11,917	2,383	434	198
Kyrgyz Republic		3062.2-4062.2[a]	510-677[a]	5,050	1,530	238	32
Romania		14,665.2[b]	2,444	200,087	64,428	No data available	No data available
Serbia	With strategic investors	3,341	557	24,270	12,017	2,898	88
	Without strategic investors	6,750	1,125				
Ukraine		30,830.4	5,138	172,830	39,887	No data available	422

Source: Data from utility companies and relevant government agencies and IMF Country Reports.

a. Options for future thermal generation include rehab of Bishkek CHP (US$350 mln) or construction of Karakeche TPP (US$ 1.2-1.5 bln).

b. Calculated based on total investment needs.

Sources of Financing Available After the Financial Crisis

The case study countries have secured less than 20 percent of the financing they will require for the investments they have planned. The financial crisis affected the availability of financing by:

- Worsening the financial performance of power sector companies, thereby diminishing their ability to fund CAPEX from their own revenues.

- Constraining the ability of commercial banks and equity investors to invest in new projects.
- Limiting the government's ability to borrow and subsidize CAPEX for publicly owned companies.[4]

Own funds

The impact of the financial crisis on power sector companies' financial performance means they have more difficulty funding CAPEX from their own revenues. Evidence of this includes the following:

- In Armenia, in generation and transmission, CAPEX from own funds decreased from 20 percent of total financing in 2006 to less than 1 percent in 2008 and only 2.5 percent in 2009. However, CAPEX from own funds is expected to increase to 4.7 percent of total financing in 2010 and 12.6 percent in 2011.
- In the Kyrgyz Republic, political uprisings in April 2010 and riots in June 2010 have left power sector companies with insufficient funds to even cover operating expenditures for the winter of 2010. Box 4.2 describes how these changes have affected the energy sector in the Kyrgyz Republic.
- In Serbia, CAPEX from own funds at EPS (state-owned generation and distribution company) are expected to decrease from an average of 76 percent (of total financing) during 2006–2008 to 36 percent from 2009–2015.
- In Romania, investments from own funds at private distribution companies declined from an average of 63 percent (of total financing) before the crisis to 39 percent in the first half of 2009. Net profit is expected to decline 59 to 75 percent at state-owned TPPs and 76 percent at Transelectrica (majority state-owned transmission company) in 2010, further reducing Transelectrica's ability to fund new investment.[5]
- In Ukraine, CAPEX from own funds at state-owned TPPs is expected to decrease from 99 percent of total financing in 2008 to only 64 percent of total financing in 2011.

Box 4.2

How will the recent political changes affect future financing of power sector investments in the Kyrgyz Republic?

In the Kyrgyz Republic, a political uprising in April 2010 and subsequent riots in June 2010 have created widespread uncertainty about future power sector investments. Key decisions made by the interim government affecting the energy sector include:

- Reversal of power and heat tariff increase implemented in January 2010.
- Reversal of the privatization of Severelectro and Vostokelectro, two of the country's four distribution companies.
- VAT and retail tax exemptions for electricity service supply.
- Maintaining social protection measures introduced in January 2010.

Key consequences of these decisions include:

- Sector cash deficit for 2010 of roughly US$ 55.6 million leaves no budget for fuel supplies required to run Bishkek and Osh CHP during the upcoming winter.
- Major cuts to capital expenditure plans in order to alleviate the state budget deficit in 2010 add to large backlog of investments creating serious risks for system reliability.

Source: Asian Development Bank. International Monetary Fund, and the World Bank. July 21, 2010. *The Kyrgyz Republic - Joint Economic Assessment: Reconciliation, Recovery, and Reconstruction.*

The ability of power sector companies to fund future CAPEX from their revenues will depend on the financial performance of these companies, which will be affected by the following factors:

- **Demand.** Revenues may increase as demand picks up in most countries in 2010.
- **Tariffs.** Tariffs will also need to increase to ensure that revenues fully cover costs—especially to cover the increased costs of imported goods resulting from currency depreciations. Governments in some countries are expected to continue postponing tariff increases throughout 2010:
  - In Armenia, the government waived return on assets for state-owned companies for 2009 and 2010, limiting future revenues available for investment.
  - In the Kyrgyz Republic, reversal of January 2010 tariff increases has created a sector cash deficit (see box 4.2).
  - In Romania, tariffs for captive residential customers will not increase until January 2011.

- In Ukraine, the moratorium on tariff increases for distribution companies has extended through 2010.
- **Operating costs.** Fuel expenditures are expected to increase further in 2010 in Armenia and Ukraine. In Armenia, many experts expect that the border price for natural gas imported from Russia will eventually reach Western European prices.[6] In Ukraine, fuel expenditures are expected to continue to increase in 2010 because the government continues to require that state-owned TPPs purchase coal from the state-owned coal mining company. Box 4.3 explains why this crisis response measure has pushed up the price of coal and negatively affected the profitability of TPPs in Ukraine in 2010.

Box 4.3

Why are fuel expenditures continuing to rise for TPPs in Ukraine in 2010?

Cabinet resolutions in October 2008, April 2009, and December 2009 required state-owned TPPs to purchase coal from SE "Coal of Ukraine" (the state-owned coal mining company) in order to support lagging demand for coal during the crisis period. Coal production at state mines nevertheless fell 15.3 percent in 2009. By the end of 2009, a recovery in steel production led a recovery in the demand for coking coal. Supply began to fall behind demand. Because of the requirement (still in place) that state-owned TPPs buy coal from state-owned mines, Ukraine has seen price increases and coal shortages.

Reserves at state-owned TPPs—especially those running on coking coal—have fallen to critically low levels. In some cases, plants have had to switch to natural gas as a fuel, further increasing costs. Burshtyn TPP, which runs on coking coal and primarily generates for the more lucrative export market, stopped exporting altogether in March 2010. Additionally, NAC ECU (state-owned company responsible for TPPs) had to take on additional short-term loans to pay for increased expenditures on coal and gas.

Box 4.3 (cont)

The combination of increased fuel and financing expenditures was expected to significantly deteriorate the financial performance of state-owned TPPs in the first quarter of 2010. Figure 4.11 shows NAC ECU's projections of profitability for 2010. NAC ECU expected profitability to improve in the second quarter of 2010 based on promises that the tariff would be reviewed on June 1, 2010.

Figure B4.3:   Projected Profitability of State-Owned TPPs in Ukraine, 2010

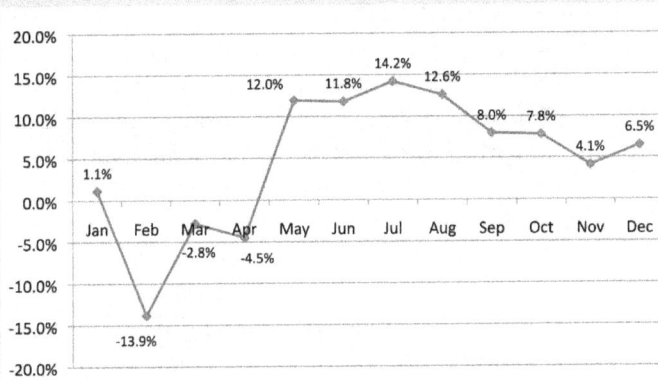

Source: NAC ECU.

The factors named above will also determine the extent to which power sector companies are able to finance CAPEX through borrowing. Deteriorating financial conditions make power sector companies less attractive for debt or equity capital. In the sections that follow we discuss the impact of the financial crisis on power sector companies' capacity to attract financing.

International Financial Institutions
In the aftermath of the financial crisis, International Financial Institutions (IFIs) will likely continue to provide most of the financing for the power sectors in the case study countries. IFIs have long been a major source of financing—especially for state-owned companies. Evidence of this can be found in each of the case study countries:

- In Armenia, funds from multilateral and bilateral IFIs accounted for 67 percent of power sector CAPEX in 2008. In 2010, funds from the IFIs are expected to account for roughly 70 percent of power sector CAPEX. For state-owned companies, IFI financing represents almost all (95%) of sector CAPEX in 2010.
- In the Kyrgyz Republic, funds from IFIs accounted for 37 percent of CAPEX for ES (the state-owned generation company) in 2006, increasing to 87 percent in 2009 with concessional financing from the Russian Government for the construction of Kambarata-2.
- In Romania, IFI financing has increased significantly in recent years. Lending from the EIB increased 30 percent when Romania joined the EU in 2007. EBRD lending to the sector increased twofold from 2008 to 2009.
- In Serbia, EPS' (state-owned generation and distribution company) financing plans indicate that concessional lending will increase from 40 percent of total financing in 2008 to 62 percent of total financing in 2015.
- In Ukraine, IFI financing at Ukrhydrenergo (state-owned HPP) increased from 0.3 percent in 2006 to 9.1 percent in 2009 under the World Bank hydropower plant rehabilitation project.

As a result of the financial crisis, private renewable energy developers in some countries are also increasingly turning to IFIs for support as other lenders have become more risk averse. As evidence of this:

- In Armenia, small hydropower (SHPP) projects have become less attractive because of increased financing costs. Some commercial banks, which committed to IFI-funded SHPPs projects, are seeking co-financing sources in AMD.
- In Romania, renewable energy project developers have increasingly turned to EBRD and International Finance Corporation (IFC) for financing because of the increased cost of commercial financing.
- In Serbia, EBRD may set up a line of credit with commercial banks in Serbia to lend for small renewable energy projects (under US$ 2 mln).
- In Ukraine, project developers are increasingly turning to EBRD and IFC because of difficulties attracting foreign equity investments.[7] The

World Bank, EBRD, and IFC are establishing a Clean Technology Fund to mobilize financing for renewable energy and energy efficiency investments by the government and private sector.

The financial crisis has not limited IFI's abilities to finance investments in the power sector, nor has it decreased power sector companies' appetite for concessional financing. However, tightened fiscal space may limit the government's ability to borrow. State budget deficits in each of the case study countries are expected to remain above precrisis levels for the next several years. Table 4.3 shows actual and projected state budget deficits estimated by the IMF for 2008 to 2011.

Table 4.3:   State Budget Deficits in the Case Study Countries, 2008–2011, % of GDP

	2008	2009	2010	2011
	Actual		Projected	
Armenia	-1.2	-8.0	-4.8	-3.9
Kyrgyz Rep	0.0	-3.7	-12	-8.5
Romania	-4.8	-7.4	-6.8	-4.4
Serbia	-2.6	-4.1	-4.8	-4.0
Ukraine	-3.2	-6.2	-5.5	-3.5

Source: IMF Country Reports.

Moreover, sovereign debt levels have increased sharply as a result of the crisis, in some cases coming close to sustainability thresholds. For example, in Serbia, 40 percent of GDP is considered the sustainability threshold for public debt. Public debt in Serbia reached 35.6 percent of GDP in 2009. Table 3.2 shows how public debt levels changed in all of the case study countries. As a result of these fiscal constraints, government's ability to borrow for power sector investments at state-owned companies will be limited.

Commercial banks

The financial crisis affected commercial lending in each of the case study countries, but the power sector remained partially insulated from these effects because there was very limited lending to the sector before the crisis. Historically, the poor financial performance of public power sector companies has limited the interest of commercial banks in the sector. Commercial banks have generally only been willing to lend to the sector for working capital needs.

In general, constraints on capital and higher country and market risk during the crisis led commercial banks to tighten lending requirements and reduce overall lending in several of the case study countries. Box 4.4 describes how the financial crisis affected commercial lending in Armenia.

Where commercial banks have provided loans to the power sector, interests rates have increased and lending conditions have tightened. For example, in Ukraine interest rates for long-term borrowings at TPPs ranged from 2.05 to 14 percent before the crisis, increasing to 19 percent during the crisis.

Box 4.4

How has the financial crisis affected commercial lending in Armenia?

A combination of higher credit risk and re-dollarization of the economy led to a decline in overall credit growth and a contraction of credit available in local currency during the crisis period in Armenia.

Higher credit risk brought on by a growth of non-performing loans led to tightened commercial lending conditions. In the first quarter of 2009, 7.8 percent of bank loans were in arrears—a two-fold increase over a six-month period. During this same period, loan/collateral ratios decreased from 60 –70 percent to 5,060 percent. Additionally, falling demand for local currency and the expected depreciation of the dram led to increased dollarization of deposits and loans at commercial banks and a resulting shortage of liquidity in local currency. Figure 4.12 shows the re-dollarization of deposits and loans at commercial banks in Armenia beginning in the fourth quarter of 2008.

Figure 4.4a: Dollarization of Loans and Deposits in Armenia

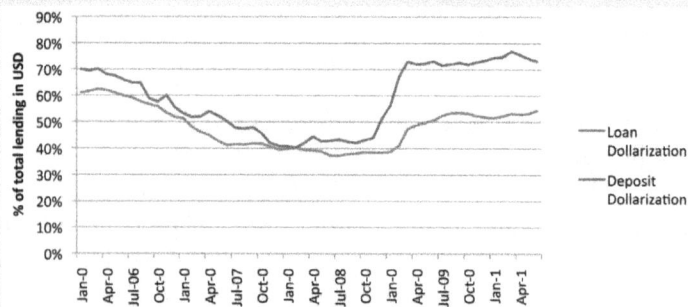

Box 4.4 (cont)

Because of the re-dollarization of the economy and tightened lending requirements, overall credit growth declined beginning in the second quarter of 2008
and loans in AMD contracted beginning in the May 2009. Figure 4.13 shows these
impacts.

Figure 4.4b: 12-Month Credit Growth in Armenia

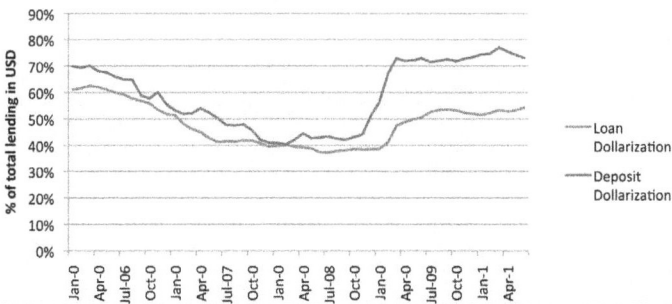

Source: IMF Armenia Team. September 9, 2009. *The Economic Crisis in Armenia: Causes, Consequences, and
Cures. Financial Banking College. Yerevan, Armenia.*

As noted above, most commercial banks' lending to the sector is for
short-term working capital requirements. Conditions for short-term loans
have also become less favorable for borrowers:

• In Armenia, interest rates for short-term borrowings without ade-
  quately liquid collateral increased from an average of 16–18 percent
  to 20–22 percent and maturities reduced from a maximum of 2.5 to
  1 year.
• In the Kyrgyz Republic, average interest rates on short-term loans for
  JSC EC (state-owned generation company) increased 2.5 percent and
  collateral requirements tightened.
• In Ukraine, working capital needs of TPPs increased significantly as a
  result of increased fuel expenditures (see box 4.3). During this period,
  interest rates increased from 19 percent to 20–26 percent.

Looking ahead, commercial banks appear to be loosening lending conditions, and credit growth is recovering. In Romania and Serbia, short- and
long-term interest rates peaked in February 2009 and have declined since.

In Armenia, loans in local currency, which contracted from December 2008 to August 2009, began to grow in the fourth quarter of 2009.

Private investors

Fiscal budgetary constraints, poor financial performance of publicly owned companies, and large investment needs have led governments in the case study countries to look increasingly to private investors to finance power sector projects.

Private sector interest has been limited, but the lack of private sector interest cannot be blamed on the financial crisis. It is generally true that foreign investors are more risk averse because of the crisis, but other factors appear to be far more important barriers to investment:

• In Armenia, feed-in tariffs are generally too low to attract private investment in renewable energy projects. Additionally, licensing and permitting processes can cause excessive delays.
• In the Kyrgyz Republic and Ukraine, privatization bids have only been able to attract local and regional bidders as the lack of transparency and need for substantial market reforms makes the sector too great a risk for most foreign investors.
• In Romania, investments in renewable energy have continued through the crisis as investors have generally considered these investments safe and highly attractive because of EU requirements and green certificate trading scheme. However, investments in conventional thermal projects have been delayed (and some have been cancelled) as investors wait to see how restructuring of generation will affect the sector. Box 4.5 describes why restructuring of publicly owned generation companies is delaying private investments in the sector.
• In Serbia, lack of consensus between government and strategic investors on a power purchase agreement and price for coal has delayed investments in two large lignite TPPs, Kolubara B and Nikola Tesla B3. Although several companies—including CEZ, Edison Italy, AES, EnBW, and RWE—expressed interest in these investments, only some have applied to continue with the selection process.

Private sector involvement in the case study countries was low before the crisis, and remains low because the country and regulatory risks remain the same. The lack of private sector financing available before the crisis is primarily attributable to poor regulatory frameworks or a failure to implement the regulatory frameworks as intended. Regulatory frame-

Box 4.5

Why is restructuring affecting private investments in generation in Romania?

In Romania, government plans to restructure the generation sector have had a major impact on the availability of financing. In 2007, the Government of Romania announced plans to organize state-owned generation plants under the ownership of one holding company. As concerns arose about the dominant position of one large company in the power sector, the government revised its plans to create two companies ("national champions").

Private investment in generation in Romania has halted since the announcement of the national champion plans. Commercial banks have postponed making any new loans to existing companies because they want to wait and see how the restructuring will affect the financial performance of the two new companies and their ability to repay debt. Foreign private investors considering Public-Private Partnerships with Termoelectrica (state-owned company of hard coal, gas, and oil fired TPPs) or investments in new greenfield capacity have postponed projects because they want to wait and see how the market share of the two new companies will affect competition and prices.

The results of this uncertainty are that:

- Many TPPs will not undergo environmental upgrades by the 2013 deadline.
- Some memoranda of understanding signed with private investors have expired and are not being renegotiated.
- Privately financed plants scheduled for 2010 will be delayed until at least 2011.

works that do not allow for full cost recovery and multi-year investment planning deter private investors from investing in new infrastructure or bidding on privatization of existing assets.

Notes

1. Appendix A provides further detail on the age, condition, and planned retirement of physical infrastructure in the power sectors in the case study countries.
2. The need for new generating capacity was estimated based on the assumption that no new capacity will be built or existing capacity rehabilitated unless financing was secured before the crisis.

3. Decline in generation and consumption primarily resulted from energy crisis. See box 1.1 for further detail.

4. A number of other factors—not linked to the financial crisis—have also affected the sector's access to financing. This section focuses solely on the impacts of the financial crisis.

5. Actions by the government of Romania in response to the crisis have also affected Transelectrica's performance. In need of additional cash, the government changed the profit payout structure for Transelectrica in 2010. Before 2010, the government received 50 percent of profits in dividends, leaving 40 percent available for reinvestment in the company (and 10 percent in bonds to employees). In 2010, the government will receive 90 percent of profits in dividends, leaving only 10 percent available for reinvestment in the company (and no profit payout to employees). Similar government plans to donate funds from the majority state-owned gas company, Romgaz, to finance the state budget deficit have been threatened with legal action by private shareholders.

6. In Armenia, gas import prices from Russia reached US$ 180/tcm in 2010. European countries imported Russian gas at nearly US$ 500/tcm in 2008. The global recession helped bring natural gas prices down to roughly US$ 325/tcm in 2010, but most experts expect a return to 2008 levels.

7. A 300 MW greenfield investment in a wind power plant in Western Crimea was delayed because the foreign equity sponsor pulled out of the project in 2009.

CHAPTER 5

Conclusions

The macroeconomic impact of the financial crisis affected the power sectors of the case study countries primarily through lower GDP, which caused lower electricity demand and hence lower revenues for many power sector companies. Currency depreciations caused higher fuel and higher debt service costs. Declining financial health—the net result of lower revenues and higher operating costs—has hurt power sector companies' abilities to fund their own CAPEX, and made it harder to raise financing and close their investment gaps. Fortunately, for many of the case study countries, the impact of the financial crisis on electricity demand delayed the need for some new investments needed to meet demand.

It is important, however, not to exaggerate the role of the financial crisis. There were and continue to be persistent, underlying policy and regulatory challenges in each country's power sector that ultimately mattered more than the financial crisis in determining capital expenditure and the availability of financing.

This section highlights the key conclusions of the report that will be most important to policymakers as they consider options for dealing with the impact of the financial crisis on their countries' power sectors. It summarizes the key impacts of the financial crisis identified earlier and describes factors that affected the power sectors in the case study countries, but were not impacts of the financial crisis (noncrisis factors). Figure 5.1 shows how the impacts of the financial crisis combined with noncrisis factors to affect the investment gaps in the case study countries.

Figure 5.1:   What were the impacts of the financial crisis?

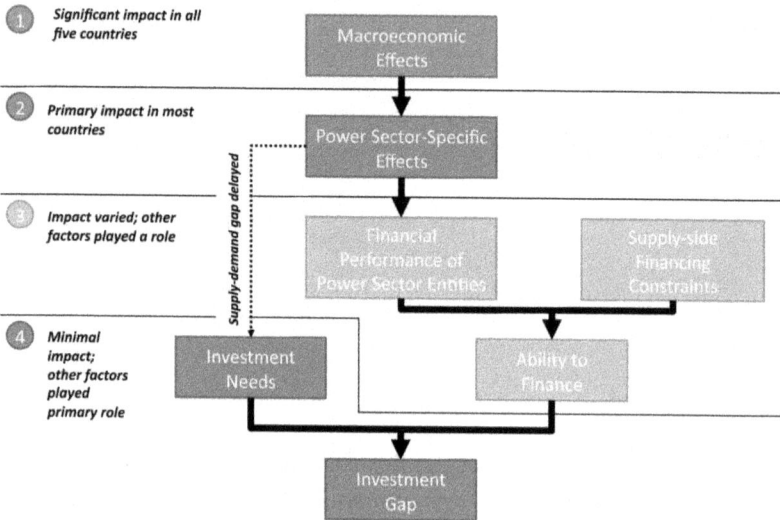

Effects of the Financial Crisis

Chapters 3 and 4 showed that the financial crisis affected the power sector in the following ways:

- The financial crisis had major macroeconomic impacts in each of the case study countries. GDP slowed or declined, currencies depreciated, and state budget deficits and public debt levels rose in each of the case study countries.
- The macroeconomic impacts had significant follow-on impacts on the power sectors of each of the case study countries. As a result of the decline in GDP, demand for electricity decreased in all countries except the Kyrgyz Republic. The drop in demand meant lower revenues but also had the effect of delaying the need for some new investment.
- Governments in all case study countries postponed tariff increases to protect certain customer groups during the crisis period.

- The net impact of lower revenues and higher operating costs affected the financial performance of power sector companies in the following ways:
  - Revenues declined in Armenia, Romania, and Ukraine as a result of the decrease in electricity demand.
  - Operating expenditures increased in Armenia, the Kyrgyz Republic, and Ukraine as the currency depreciation resulted in increased costs for imported fuel and higher debt service costs.

The financial crisis also affected, to some extent, the availability of future financing because of:

- The poor financial performance of some power sector companies during the crisis period, which will limit their abilities to fund CAPEX from own funds.
- Supply-side constraints that limit commercial bank lending and the availability of capital for equity investments.
- Fiscal constraints on governments' borrowing capacities.

Table 5.1 summarizes the potential for meeting CAPEX needs with various sources of financing. As noted in chapter 4, the availability of private financing was limited before the financial crisis, and remains limited for reasons related more to the underlying investment conditions in each country's power sector, than the financial crisis.

What the Financial Crisis Did Not Affect

It is informative also to recognize what the financial crisis did not affect in the power sectors of the case study countries. The financial crisis had little effect on:

- **Capital expenditures and investment planning.** Changes in capital expenditure depended primarily on factors that existed before the crisis or that coincided with the crisis, namely:
  - The regulatory environment. Changes in the level of capital expenditure during the crisis depended on whether the regulatory environment allowed for recovery of CAPEX and return on investment.
  - Other factors that coincided with the crisis, including the political crisis in the the Kyrgyz Republic, the plan to restructure generation ownership in Romania, and the need (in Serbia, Romania and Ukraine) to comply with EU environmental regulations.

Table 5.1:   Likelihood of Increased Financing from Various Sources After the Crisis

	Likelihood of increased financing	Reasons why financing likely/unlikely to increase post-crisis	Examples
Own funds	No	Financial crisis has negatively affected financial performance of some power sector companies, and government reactions to crisis have further hurt performance.	• Ukraine TPPs are expected to show negative net income for first 3 months of 2010 • Private distribution companies in Romania may cut CAPEX plans if revenue "true-up" further postponed
IFIs	Depends	IFIs have been and continue to be the primary source of financing for the sector in most countries, but increased financing may be limited. Government fiscal space is limited, so may struggle to take on additional loans from IFIs.	• IFI funds as % of secured financing: Armenia: 81%; Kyrgyz Republic: 100% • Serbia: 100% (transmission) • Ukraine: 88% (transmission); 82% (HPPs); 40% (TPPs)
Commercial lenders	No	Commercial lending has primarily been used only for working capital, but interest rates have increased, maturities shortened, and collateral requirements tightened.	• Average interest rates on short-term loans to power sector companies rose 4 –6% in Armenia, 2.5% in Kyrgyz Republic, and as much as 6% in Ukraine
Private investors	Yes	Private investors are generally more risk averse as a result of the crisis, but are largely influenced by other factors, which, if addressed, can increase potential for private sector participation.	• Private investment limited by regulatory environment in Kyrgyz Republic and Ukraine pre- and postcrisis • Affected by other factors in Romania (restructuring) and Serbia (negotiations of PPA and price of coal)

Source: Authors.
Note: IFI = International Financial Institutions.

The investment gap was wide before the crisis and remains large after it, despite the drop in electricity demand.

- **The availability of private financing.** Commercial bank financing and private investment in the case study countries was scarce before the crisis and continues to be in 2011. The cause is not the financial crisis, but instead a variety of country and regulatory risks and (as a consequence, in part, of the former) the historically poor financial performance of public power companies.

It is also informative to look at the differences among the case study countries to understand what affected CAPEX and the availability of financing in each. Power sector companies in three of the case study countries (the Kyrgyz Republic, Serbia, and Ukraine) had chronic difficulties meeting their investment needs before the crisis. Private sector participation was largely absent from their power sectors, and commercial lending was limited to meeting working capital needs, just as it is now. Investment plans were rarely met. In contrast, power sector companies in Armenia and distribution and transmission companies in Romania more regularly met their investment plans before the crisis, and continue to do so, though they may have scaled back those plans in response to the crisis.

CHAPTER 6

Recommendations

The financial crisis makes clear the importance of the following:

- **Prioritization of public spending.** With smaller public budgets and scarcer commercial lending, governments will need to prioritize power sector investments carefully. In all of the case study countries, energy efficiency is a least cost solution that can postpone the emerging supply-demand gap. Governments will also need to carefully balance capital expenditures—taking into consideration life-cycle investment costs—with operating and maintenance expenditures as some operating expenditures, particularly fuel costs, continue to grow. This includes considering tradeoffs between new investment and expenditure on maintenance needed to preserve existing infrastructure

- **Creation of a more attractive environment for investment.** Power sector companies must become financially viable in order to attract financing for needed investments. Policymakers can help create a financially viable power sector through policy, legal, institutional, and regulatory reform. Power sectors in the case study countries can benefit from the creation and implementation of laws and regulations that support the enforcement of contracts and property rights and allow for full cost recovery and predictable recovery of capital expenditure.

The World Bank can assist governments in implementing both of these recommendations through a combination of loans, guarantees, and technical assistance. The following subsections outline each of the recommendations and the role for the World Bank in more detail by proposing a prioritization of public spending in each of the case study countries' power sectors,[1] describing the changes that governments can make to better attract private sector investment, and suggesting possible roles for the World Bank in helping to implement the recommendations.

Prioritize Public Spending

Growing investment needs and limited financing make the prioritization of power sector investments extremely important. This is especially true in the wake of the financial crisis. Governments will need to consider the implications of new power sector investments in terms of affordability, supply reliability, and energy security. In Romania, Serbia, and Ukraine, the governments will also need to prioritize the investments required for compliance with certain EU regulations. Table 6.1 shows a prioritization of short- and long-term investments in each of the case study countries based on criteria of supply reliability, affordability, and compliance with EU regulations.[2] Some of the case study countries require immediate investment. the Kyrgyz Republic, for example, currently faces the threat of a winter energy shortage because of insufficient baseload capacity. Others should begin making investments incrementally now to avoid severe consequences in the next few years. For example, Romania will need to shut down a significant portion of its existing capacity, or pay large fines to keep it operational, if it does not invest in environmental upgrade of its TPPs.

Priorities within each of these criteria differ for each of the case study countries. Unfortunately, data were not available to evaluate investments in each country for each of the criteria. Table 6.2 and table 6.3 show which criteria we used to rank investments in each country in generation and transmission, respectively.

Table 6.1:   Short- and Long-term Priority Investments in Each Country

	Short-term (1–3 years)	Long-term (4–7 years)
Armenia	Transmission rehabilitation	Construction of new NPP & RE capacity
Kyrgyz Republic	Urgent rehabilitation to improve baseload capacity for upcoming winter	Transmission rehabilitation
Romania	Environmental upgrades of TPPs; relaunching nuclear company; distribution rehabilitation; transmission connections for RE and interconnections	New capacity (conventional thermal, nuclear, hydro, and wind), transmission and distribution rehabilitation
Serbia	Environmental upgrades of TPPs, transmission interconnections, and distribution rehabilitation	New capacity, transmission and distribution rehabilitation
Ukraine	Rehabilitation of HPPs; rehabilitation of TPPs	Service life extension of NPPs

Source: Authors.

Note: These priority investments are based on the criteria and methodology described in further detail in appendix B and do not reflect the World Bank's investment strategy in the case study countries.

Table 6.2:   General Priorities for Generation in the Case Study Countries

	Supply reliability		Affordability	EU regulations
	Adequacy	Security	Affordability	EU regulations
Armenia	Baseload capacity	Uses domestic resources	Lowest levelized cost	
Kyrgyz Republic	Winter baseload capacity	Uses domestic resources and increases supply diversity	Lowest unit cost	
Romania	Baseload capacity	Uses domestic lignite and uranium		Complies with EU emissions and RE regulations
Serbia	Short-term: Rehabilitate peak capacity Medium-term: New baseload capacity			Complies with EU emissions regulations
Ukraine	Baseload capacity	Uses domestic resources	Lowest levelized cost	

Source: Authors.

Table 6.3:   General Priorities for Transmission in the Case Study Countries

| | Supply reliability | | | |
	Adequacy	Security	Affordability	EU regulations
Armenia	Oldest, greatest number of outages, longest outage duration			
Kyrgyz Republic	Greatest number of customers affected		Lowest total cost of investments	
Romania	1st priority: Improving reliability of substations and 220 kV lines	2nd priority: Improving interconnections	1st priority: Reducing O&M costs	2nd priority: Connecting RE capacity
Serbia	System technical requirements; Assets in poorest conditions			
Ukraine	Greatest number of avoided losses and reduction in energy not served	Increased import capacity		

Source: Authors.

The tables above are based on an indicative prioritization framework developed for this report; they are not a substitute for a detailed power sector planning exercise. The tables can, however, provide the basis for a discussion about the hard choices that will need to be made between investments for which limited public funding is available.

Rational CAPEX planning is especially important where power sector companies are mostly publicly owned. For well-run, publicly owned power sector companies, the planning process begins with least cost sector development plans. For regulated markets, these physical plans are then integrated with multiyear financing plans that are approved by the regulator and fully reflected in the tariff. As mentioned above, although most power sector companies in Armenia have needed to scale back their investment targets in recent years, they have largely managed to meet them because the regulatory regime allows for multiyear investment planning and predictable recovery of investment costs. This is also true in Romania where privatization of five distribution companies depended, in part, on the credibility of sector regulation to create and implement a tariff methodology that allowed for recovery of investment costs.[3]

In the other case study countries, investment plans typically far exceed what is possible given the funds available because of problems with the regulatory frameworks or because of failure to apply the frameworks as they were intended. The quality of power sector planning is, in part, a function of the incentives provided by public owners or sector regulators. Power sector companies in the Kyrgyz Republic, Serbia, and Ukraine face tariffs that are generally below the cost of service, leaving little money for debt service once operating and maintenance costs are paid. The companies also face investment approval processes that are unpredictable, ad hoc, and often driven more by political than commercial and technical considerations.

Create Favorable Environments for Investment

Power sector companies must become financially viable in order to attract financing for needed investments. Private companies will invest in electricity sectors where they think they will be able to earn enough revenue to cover their operating and maintenance costs, service their debt, and pay the level of returns expected by shareholders. They will generally be willing to take operational and commercial risks associated with generating or distributing electricity, but will not take risks that their revenues will be disrupted by political changes or changes to the way in which their tariffs are determined. The same is true for financiers of public companies. Commercial and IFI lending to public companies may also disappear if it becomes clear that the public companies will have difficulty servicing their loans.

A policy, legal, and regulatory environment that supports a financially viable sector is essential for attracting private investors. The case studies in this report and in earlier World Bank reports strongly support this conclusion.[4] Important specific ingredients in such an environment include:[5]

- Laws and regulations that support the enforcement of contracts and property rights, including the disconnection of nonpaying customers and punishment for electricity theft. This enforcement is essential to safeguarding power sector companies' cash flows.
- Regulation that allows for full cost recovery of reasonable capital expenditure and the costs (debt service or dividend payments) required to finance it. This is essential to ensuring that companies in the power sector generate enough internal cash for operations and maintenance, debt service, and any equity contribution to capital expenditure
- Regulations that allow for predictable regulatory approval of the costs of investment plans. Power sector companies make more realis-

tic investment plans—in other words, plans that they are able to implement—if they understand the criteria by which those plans are evaluated and believe they understand how the criteria will be applied. A predictable investment approval process will balance criteria of affordability against the need for improvements in quality and reliability of service. This, in turn, requires: (i) clear targets that reflect customer preferences regarding service quality and reliability and (ii) knowledge of what customers are actually able to afford.

Box 6.1 provides a list of 10 rules, identified in the World Bank's energy flagship report for the ECA region, which can further help foster an investment climate that attracts private sector participation.

A comparison of privatization efforts in the case study countries confirms these lessons. Evidence from the Kyrgyz Republic and Ukraine in

Box 6.1

Seven Do's and Three Don'ts for Creating a Better Investment Climate

The World Bank's energy flagship report for Eastern Europe and Central Asia (ECA) titled, *Lights Out? The Outlook for Energy in Eastern Europe and Central Asia*, (see box 2.1) identified 10 rules to follow to help improve the investment climate in the region:

1. Don't impose a punitive or regressive tax regime.
2. Do introduce an acceptable legal framework.
3. Do provide supporting regulations administered by an independent and impartial regulator.
4. Do create an environment that facilitates assured nondiscriminatory access to markets.
5. Don't interfere with the functioning of the marketplace.
6. Don't discriminate among investors.
7. Do honor internationally accepted standards.
8. Do abide by contractual undertakings and preclude the use of an administrative bureaucracy to constrain investor activities.
9. Do prevent monopoly abuses.
10. Do ensure that the sector is kept free of corruption.

Source: World Bank. 2010. *Lights Out? The Outlook for Energy in Eastern Europe and the Former Soviet Union.* Washington, DC: World Bank.

particular suggest that efforts to privatize without these ingredients often end in failure. Romania and Armenia, in contrast, are the case study countries with the most successful records of private investment in electricity because sector regulation has ensured that investors will recover their investment costs.

Good governance is an important determinant of private sector participation, where governance encompasses a range of characteristics, including rule of law, regulation, control of corruption, government effectiveness, and transparency. Among the study countries, Romania and Armenia rank higher relative to most key governance indicators tracked by the World Bank's Worldwide Governance Indicators (WGI) Project (see figure 6.1).[6]

Figure 6.1:   World Bank Governance Indicators for the Five Case Study Countries

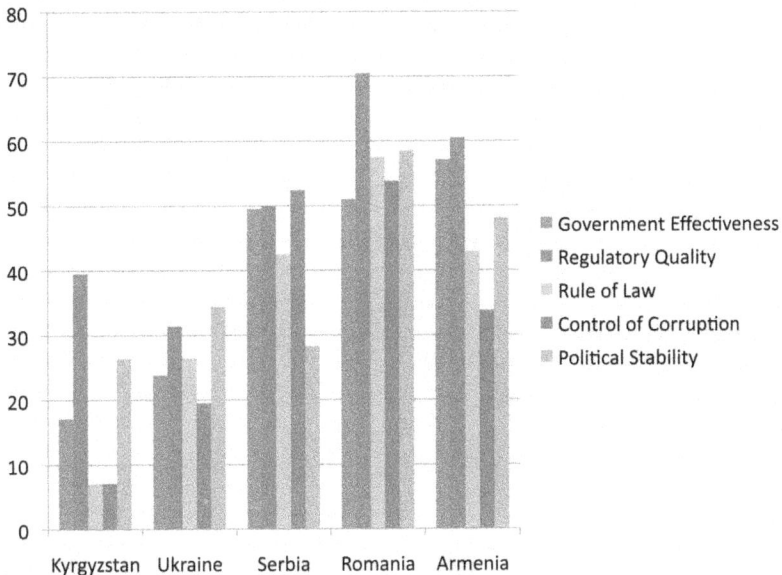

Source: D. Kaufmann, A. Kraay, and M. Mastruzzi (2010), *The Worldwide Governance Indicators: Methodology and Analytical Issues.*

Role for the World Bank

The World Bank can help the case study countries emerge from the financial crisis by supporting public spending on critical power sector investments, and supporting government efforts to create environments conducive to investment in the sector. The World Bank can do this by providing:

- Loans for physical infrastructure,
- Advisory services and technical support, and
- Guarantee instruments.

The following sections outline how the World Bank can support the governments in each of these areas.

Loans for physical infrastructure

The World Bank can provide financial assistance by financing physical infrastructure. This includes some of the priority investments identified earlier in this chapter and in appendix B. Box 6.2 describes World Bank financing of physical infrastructure that has supported the creation of a regional market for electricity in the countries of South East Europe.

Box 6.2

Creating Regional Markets for Electricity in South East Europe

The conflict of the 1990s divided the once unified electricity system of the countries of South East Europe (SEE) into several systems. Transmission interconnections were destroyed. Meanwhile, growing demand and limited supply have threatened to hamper economic activity in many countries.

In recognizing the value of regional cooperation to address their energy concerns, governments of the SEE countries created the Energy Community of South East Europe (ECSEE). The aim of ECSEE is to:

- Rebuild the region's energy networks,
- Create a stable climate to foster investment in the sector, and
- Establish conditions in which economies can be rebuilt effectively.

Rebuilding the region's transmission infrastructure to support a strong regional market, which further fosters investment, is a vital piece of the ECSEE framework. The World Bank's ECSEE Adaptable Program Loan (APL) series has supported efforts in seven countries through ten loans and credits to rehabilitate and

Box 6.2 (cont)

expand their transmission networks to support regional trade. Several examples include:

- In Macedonia, the World Bank provided US$ 25 million to expand the Skopje substation and upgrade, rehabilitate, and construct various interconnections, overhead lines, and 110 kV substations.
- In Turkey, the World Bank provided US$ 66 million under Turkey's APL 2 for implementation of a market management system (MMS), a national load dispatch center (SCADA/EMS), and transmission system reinforcement, including rehabilitation of substations.
- In Serbia, the World Bank provided US$ 21 million for the construction of two new 110 kV substations and new 110 kV interconnection lines for these substations.

Sources: World Bank. March 14, 2005. Energy Community of South East Europe (APL #2) (Turkey). Project Appraisal Document. Washington, DC: World Bank; World Bank. December 8, 2005. Federal Yugoslav Republic of Macedonia - Third Energy Community of South East Europe Program (ECSEE APL 3) Project. Project Appraisal Document. Washington, DC: World Bank; World Bank. June 8, 2005. Energy Community of South East Europe (APL) Program - Serbia and Montenegro Component - Serbia Project. Project Appraisal Document. Washington, DC: World Bank.

Advisory services and technical support

The World Bank can offer advisory services that support government efforts to:

- **Improve public sector investment planning.** The World Bank can provide support to overall energy sector planning and the development of Least Cost Development Plans (LCDPs). LCDPs are especially needed in Ukraine and the Kyrgyz Republic as a baseline for investment planning. Armenia could also benefit from an updated LCDP given the impact of financial crisis on demand, the rising cost of nuclear plant construction, and increasing natural gas prices.
- **Enact regulatory and market reform.** The World Bank has extensive experience helping governments enact regulatory reforms that promote financially viable power sectors while designing subsidy schemes that protect vulnerable customers. Such reforms are especially needed in the Kyrgyz Republic, Serbia, and Ukraine.

Box 6.3 describes how World Bank engagement in the energy sector has helped Turkey on its path toward EU accession by fostering electricity market reforms and promoting private sector participation in clean energy.

Box 6.3

World Bank Assistance in Support of Turkey's Energy Reforms

Turkey's energy sector is a critical component of its EU accession process. To this end, the government has embarked on an impressive shift toward clean energy and an opening of its electricity market. Meanwhile, the government has sought to balance its EU environmental and competition requirements while maintaining commitments to energy security and sustainable growth in the sector.

The World Bank's engagement in the energy sector in Turkey through its Environmental Sustainability and Energy Sector DPL 2 loan has supported the government's multitiered strategy. Specifically, the World Bank has helped the Government of Turkey to:

- Implement market reform. With World Bank support, the government implemented a cost-based electricity pricing scheme and electricity market regulation, paid arrears to electricity suppliers, privatized seven distribution companies, and prepared for generation privatization.
- Rehabilitate transmission and distribution networks. Transmission capacity increased substantially and distribution reliability improved. From 2002 to 2009, electricity transmission capacity increased 70 percent. From 2004 to 2008, supply reliability almost doubled.
- Support clean energy initiatives. Electricity produced from privately owned renewable generation facilities more than doubled over the course of the World Bank DPL 2 loan. Additionally, the government recently finalized its national Climate Change Strategy.

Following the success of the World Bank's long-term engagement in the energy sector, Turkey recently received US$ 600 million from the World Bank for its Private Sector Renewable Energy and Energy Efficiency Project to further mobilize private financing of clean energy projects. The project will be the first to receive funding (US$ 100 million) from the new Clean Technology Fund managed by the World Bank.

Sources: World Bank. October 5, 2010. Turkey Second Environmental Sustainability and Energy Sector Development Policy Loan (ESES-DPL 2). Washington, DC: World Bank; World Bank Press Release. "Turkey Receives World Bank and First-Ever Clean Technology Fund Financing for Renewable Energy and Energy Efficiency Program." 2009/ECA/368.

Partial Risk and Partial Credit Guarantees

World Bank Partial Risk Guarantees (PRGs) and Partial Credit Guarantees (PCGs) can lower the borrowing costs of private investors in the power sector.

PCGs insure commercial lenders against default by public companies or government agencies, thereby lowering the cost of commercial debt available to these companies.

World Bank PRGs can help reduce political and regulatory risk for private investors seeking to privatize or enter into Public Private Partnership (PPP) arrangements. PRGs may be a useful way to further incentivize private sector participation after other regulatory and market reforms have taken place. The World Bank's success of providing a PRG to ENEL in the privatization of two of Romania's eight distribution companies incentivized privatization of three other distribution companies in Romania as these companies indirectly benefited from the regulation backstopped by the PRG. Based on the successful experience of the PRG in Romania, CEZ made a similar PRG a precondition to its investment in Albania. Box 6.4 describes how a PRG in Albania contributed to the successful privatization of a distribution company to CEZ.

Box 6.4

How a PRG Enabled Privatization of a Distribution Company in Albania

The Government in Albania wanted to privatize the Energy Distribution System Operator of Albania (OSSH) in 2006, but potential investors expressed concerns. Specifically, they were concerned with the new regulatory framework implemented by a regulatory agency with a limited track record, the projected tariff adjustments needed to bring tariffs to cost-recovery levels, and the upcoming elections, which posed critical regulatory risks.

The World Bank's track record in the sector made it well-positioned to help mitigate potential regulatory risks. In 2008, through a competitive bidding process, CEZ a.s. purchased OSSH for €102 million. The share purchase agreement (SPA), under which CEZ purchased 76 percent of the share capital of OSSH, ensured a balanced approach under which the Government of Albania and the regulatory agency agreed to a regulatory framework that would provide reasonable returns to OSSH once pre-agreed performance targets had been reached.

Box 6.4 (cont)

A Partial Risk Guarantee (PRG) provided by the World Bank was a precondition to the two parties signing the SPA, which led to financial closure of the privatization. The PRG guaranteed the government's obligation to compensate the privatized OSSH if the regulatory agency or the government failed to implement the regulatory framework agreed to under the SPA.

Along with being critical to the financial closure of the privatization, the PRG strengthened the sector by the following:

- Helping to attract an important regional player to the sector,
- Reinforcing the independence of the sector regulator, and
- Helping to mobilize up to €240 million in OSSH in the next five years.

Source: World Bank. May 2009. "Financial Solutions: Partial Risk Guarantee - World Bank Issues Regulatory Risk Guarantee in Support of Albania Electricity Sector" Retrieved November 11, 2010 from: http://sitere-sources.worldbank.org/INTGUARANTEES/Resources/AlbaniaOSSHElectricityPrivatization.pdf.

Notes

1. Appendix B describes in more detail the methodology used for prioritizing investments and shows the resulting prioritization of specific investments within each segment (generation, transmission, and distribution).

2. The prioritization only includes investments that have not yet secured financing and are likely to receive partial or full public (government) funding.

3. Recent government actions, however, have undermined the credibility and independence of regulation in Romania, threatening to affect future investment planning in the sector.

4. See Lights Out? The Outlook for Energy in Eastern Europe and Central Asia and Venkataraman Krishnaswamy, and Gary Stuggins, Closing the Electricity Supply-Demand Gap. Energy and Mining Sector Board Discussion Paper, Paper No. 20. Washington, DC: World Bank, January 2007.

5. Implementation of regulation is as important as its design. A good regulatory framework on paper will not attract investment if it is not implemented, or if government interferes in its implementation.

6. The indicator for "voice and accountability" is not shown here.

Overview of Power Sectors in Case Study Countries

This appendix provides a brief overview of the power sectors in each of the case study countries. The following sections outline some key statistics for each case study on sector structure, age and condition of assets, capacity, generation, consumption, and tariffs.

Sector Structure and Main Entities

This section describes the sector structure and main entities operating in each of the case study countries. Table A.1 includes information on the structure of the power sector in each of the case study countries, including whether the sector is unbundled, the name of the regulator, the level of private sector investment, the policymaking body, and the market structure. Table A.2 lists the main companies operating in segment—generation, transmission, and distribution—of the power sector.

Table A.1:   Power Sector Structure in the Case Study Countries

	Unbundled	Regulator	Private investment			Market structure	Policymaking entity
			G	T	D		
Armenia	Y	Public Services Regulatory Commission	52%	0%	100%	Single-buyer model	Ministry of Energy and Natural Resources
Kyrgyz Republic	Y	Ministry of Energy	7%	7%	7%	Bilateral contract model	Ministry of Energy
Romania	Y	National Agency for Energy Regulation	1%	0%	37.5%	Competitive wholesale market	Ministry of Economy
Serbia	Y	Agency for Energy of Republic of Serbia	0%	0%	0%	Fully regulated[a]	Ministry of Mining and Energy
Ukraine	Y	National Electricity Regulatory Council	14%	0%	39%	Hybrid arrangement[b]	Cabinet of Ministers

Source: Data from utility companies and relevant government agencies.

a. Only wholesale market for export and import is competitive.

b. Market in transition to wholesale market with bilateral contracts.

Table A.2:   Main Power Sector Companies in the Case Study Countries

	Segment	Company	Description
Armenia	Generation	Vorotan HPP	State-owned hydropower plant
		Sevan-Hrazdan HPP	Privately owned hydropower plant
		ANPP "Armenia Nuclear Power Plant"	Publicly owned nuclear plant
		Hrazdan TPP	Privately owned thermal power
		Yerevan TPP	State-owned thermal power plant
	Transmission	HVEN "High Voltage Electricity Network of Armenia"	State-owned transmission system operator
	Distribution	ENA "Electricity Network of Armenia"	Privately owned distribution company
Kyrgyz Republic	Generation	JSC ES "Electrical Stations"	State-owned generation company
	Transmission	JSC NESK "National Electrical Grid of Kyrgyz Republic"	State-owned transmission company

	Segment	Company	Description
Kyrgyz Republic	Distribution	JSC Severelectro	State-owned distribution companies serving four different regions of Kyrgyz Republic
		JSC Vostokelectro	
		JSC Oshelectro	
		JSC Jalabatelectro	
Romania	Generation	Rovinari	State-owned lignite thermal power plants
		Craiova	
		Turceni	
		Termoelectrica	State-owned company owning hard coal, and gas- and oil-fired thermal power plants including: ELCEN, Galati, Deva, Borzesti, Doicesti, Braila, and Paroseni
		Nuclearelectrica	State-owned company owning Cernavoda nuclear power plant
		Hidroelectrica	State-owned hydropower company
	Transmission	Transelectrica	Majority state-owned transmission system operator
	Market Operator	OPCOM	Wholesale market operator
	Distribution	Electrica	State-owned distribution company owning three distribution companies in Romania including: Electrica Muntenia Nord, Electrica Transylvania Nord, and Electrica Transylvania Sud
		ENEL	Privately owned distribution company owning three distribution companies including: Electrica Dobrogea, Electrica Banat, and Electrica Muntenia Sud
		E.ON	Privately owned distribution company - Electrica Moldova
		CEZ	Privately owned distribution company - Electrica Oltenia
Serbia	Generation Distribution	EPS "Electric Power Utility of Serbia"	State-owned vertically integrated company owning coal production, electricity generation, and electricity distribution
	Transmission	EMS "Elektromreža Srbije"	Serbian transmission system and market operator

	Segment	Company	Description
Ukraine	Generation	NAC ECU "Energy Company of Ukraine"	State-owned holding company for thermal power plants, large combined heat and power plants, and public distribution companies
		Energoatom	State-owned nuclear power company
		UHE "Ukrhydrenergo"	State-owned hydropower company (ownership recently transferred to NAC ECU)
		DTEK	Privately owned company owning coal production and thermal power plants
	Transmission	Ukrenergo	Transmission system operator
	Distribution	NAC ECU	See above
		Multiple private owners	39% of distribution is privately owned

Source: Data from utility companies and relevant government agencies.

Age and Conditions of Physical Infrastructure

The power sector in each of the case study countries is characterized by old, Soviet-era infrastructure. Table A.3 describes the average age and conditions of generation, transmission, and distribution infrastructure assets in each of the case study countries.

Table A.3:   Age and Condition of Power Sector Infrastructure in the Case Study Countries

Infrastructure assets		Average age (yrs.)	Condition of assets
Armenia	Generation	40%> 40	Residual life: S-H Cascade≈5.4-10.8 yrs.; Yerevan TPP≈2.4-3.2 yrs.; Hrazdan TPP≈12.7 yrs.; ANPP≈8.8 yrs.
	Transmission	45	20% of 220 kV lines (~300km) require urgent rehabilitation and modernization
	Distribution	32	42% of substations in very poor technical condition; 14,000 autotransformers under- or overloaded
Kyrgyz Republic	Generation	32	All 16 HPPs in need of significant rehabilitation
	Transmission	34	~20% of transmission lines >40 yrs. old
	Distribution	31	17% of distribution lines in unsatisfactory conditions; 8% are unserviceable

Infrastructure assets		Average age (yrs.)	Condition of assets
Romania	Generation	33.4	TPP: 80% exceeded design life, most require environmental upgrades; HPPs: 37% exceeded design life; NPPs: good condition
	Transmission	Unknown	In good condition due to major investment program (~1 bln EUR) undertaken in recent years
	Distribution	33.7	65% of distribution networks marked by extensive use, contributing to annual losses of 12.6%
Serbia	Generation	33.6	Coal-fired TPPs and gas-fired CHPs inefficient and in need of reconstruction; revitalization needed for most HPPs before 2015
	Transmission	20.9	Poor/Adequate/Good: Substations = 46%/35%/19%; OHLs=28%/46%/26%; Pylons=28%/46%/26%
	Distribution	Unknown	Improvement of metering and rehab of existing facilities needed
Ukraine	Generation[a]	37	Most TPPs have exceeded their design life
	Transmission[b]	>25	Some equipment depreciated and uses outdated technology
	Distribution[b]	Unknown	17% of distribution lines in poor technical condition; 13% of transformers exhausted service life

Source: Data from utility companies and relevant government agencies.

a. Generation ownership split as a percentage of installed capacity (MW).

b. Transmission and distribution ownership split as a percentage of power supplied (kWh).

Large amounts of generation capacity must be replaced or rehabilitated in the case study countries because of years of under-maintenance or because they have reached the end of their design life:

- In Armenia, 1,257.5 MW—including Armenia Nuclear Power Plant (ANPP) (407.5 MW), Hrazdan Thermal Power Plant (TPP) (800 MW), and Yerevan TPP (50 MW)—must be retired by 2017.
- In the Kyrgyz Republic, Bishkek Combined Heat and Power Plant (CHP) and Uch-Kurgan Hydro Power Plant (HPP) require rehabilitation to increase their combined capacity from 275 MW to 530 MW.
- In Serbia, EPS plans to retire 2,984 MW by 2021:
  - 2015: CHP Novi Sad (208 MW)
  - 2016: TPP Kostolac B (640 MW)
  - 2017: TPP Morava (108 MW)
  - 2018: TPP Kostolac A (281 MW)
  - 2019: TPP Kolubara (245 MW)
  - 2021: TPP Nikola Tesla A (1502 MW)

- In Ukraine, 10,300 MW of TPPs require rehabilitation and 10,318 MW of NPPs require service life extension before 2020.

Snapshot of Key Statistics before the Financial Crisis

Table A.4 provides a snapshot of key statistics before the financial crisis—including installed and available capacity, generation, consumption, transmission and distribution losses, and the generation mix. Statistics are provided for 2008.

Table A.4:   Snapshot of Key Sector Statistics Before the Crisis

	Armenia	Kyrgyz Republic	Romania	Serbia	Ukraine
Installed capacity	3,655 MW	3,680 MW	20,380 MW	7,591 MW	49,267 MW
Available capacity	1,466 MW	3,135 MW	13,298 MW	7,119 MW	41,534 MW
Peak demand	1,190 MW	2,970 MW	9,369 MW	6,383 MW	30,079 MW (Winter)
Consumption	4,379 GWh	7,016 GWh	48,672 GWh	33,292 GWh	144,874 GWh,
T&D losses	13.9%	34%	12.6% (distrib.)	18.5%	14%
Generation mix	Nuclear: 44% Thermal: 29% Hydro: 26.95% Other: 0.05%	Hydro: 90% Thermal: 10%	Thermal: 54.1% Hydro: 28.4% Nuclear: 17.5%	Thermal: 70% Hydro: 30%	Nuclear: 47% Thermal: 43% Hydro: 6% Other: 4%

Source: Data from utility companies and relevant government agencies.

Tariffs

Tariffs in the case study countries generally do not cover long-run marginal costs. Table A.5 indicates when tariffs were most recently revised and the average residential tariff in 2008.

Table A.5:   Tariffs in the Case Study Countries

	Most recent tariff revision?	*Residential tariffs (US cents/kWh)*
Armenia	G/T/D: 1 Mar 2010 End-users: 1 Apr 2009	7.92
Kyrgyz Republic	New mid-term policy adopted in 2009 led to two-fold tariff increase on 1 Jan 2010, but tariff increase has since been reversed	1.72
Romania	2008	14.5
Serbia	1 Mar 2010	6.51
Ukraine	1 Sep 2006	2.85-3.85

Source: Data from utility companies and relevant government agencies.

Priority Investments in Case Study Countries

In this appendix, we describe the criteria and methodology used for prioritizing investments for each segment of the power sector: generation, transmission, and distribution, where possible. We rank only those investments that do not have financing secured and are expected to receive public funding, for example, from the state budget or sovereign guaranteed borrowing. Generally, within each segment we use two criteria to prioritize investments:

- Supply reliability. We assess supply reliability based on whether a particular investment provides supply adequacy and supply reliability, using standard industry definitions of those concepts:
  - Adequacy. "The ability of the power system to supply the aggregate electrical demand and energy requirements of customers at all times, taking into account schedule and reasonably expected unscheduled outages of system elements."
  - Security. "The ability of the power system to withstand sudden disturbances such as electric short circuits or unanticipated loss of system elements." We consider fuel supply under "system elements" and so include fuel supply security in this definition.

- Affordability. We assess affordability based on the impact of investments on end-user tariffs.

The following sections describe for each country how we measure and rank supply reliability and affordability within each segment, and how we develop an overall prioritization rank based on the two criteria. Our measure of supply reliability and affordability differs from country to country depending on what data were available when the study was conducted.

Armenia

The following subsections describe how we prioritize generation and grid development investments in Armenia.

Generation

We assess supply reliability of generation in Armenia based on the following parameters:
- Which type of capacity is most needed in order to continue to reliably serve daily load?
- Does the investment increase fuel supply security in Armenia?

We rank each of these parameters based on the following factors:
- Role in meeting daily load. We rank the role that each type of generation plays in meeting daily load as follows:
  - Baseload = 1 (highest priority)
  - Peak = 2
  - Not dispatchable = 3 (lowest priority)

  Justification: Armenia will need a new large source of baseload capacity when the Metsamor nuclear plant is decommissioned in 2016.
- Security of supply. We rank security of supply as follows:
  - Uses domestic resources = 1
  - Uses imported fuel supply = 2
  - Will be used primarily for export purposes = 3

We rank each generation investment against each of these factors and sum the rankings to arrive at a total rank for supply reliability. We then assign each investment a corresponding high (1), medium (2), or low (3) rank to allow for comparison with the affordability rank order. Table B.1 demonstrates how this is done for the generation investments being considered in Armenia.

Table B.1:   Supply Reliability Rank for Generation in Armenia

Investment description	Role in meeting daily load	Security of fuel supply	Total supply reliability rank	Corresponding rank
Vorotan HPP	2	1	3	1
Energy efficiency	1	1	2	1
Sevan-Hrazdan Cascade	3	1	4	2
Shnokh HPP	2	1	3	1
Lori-Berd HPP	2	1	3	1
Small Hydro	3	1	4	2
Meghri HPP	3	3	6	3
Pumped storage	2	1	3	1
Replacement of ANPP	1	2	3	1
Wind power	3	1	4	2

Source: Authors.

We assess affordability as the impact on end-user tariffs by comparing the levelized cost (LEC) of each investment. Levelized costs are ranked as follows:

- Less than US$ 0.05/kWh = 1
- Between US$ 0.05/kWh and US$ 0.10/kWh = 2
- Greater than US$ 0.10/kWh = 3

Table B.2 shows the levelized costs of each investment and the investment's corresponding affordability rank.

Table B.2:   Affordability Rank for Generation in Armenia

Investment description	Levelized energy cost (US$/kWh)	Corresponding rank
Vorotan HPP	0.013	1
Energy efficiency	0.015	1
Sevan-Hrazdan Cascade	0.013	1
Shnokh HPP	0.074	2
Lori-Berd HPP	0.096	2
Small Hydro	0.042	1
Meghri HPP	0.035	1
Pumped storage	No data	3
Replacement of ANPP	0.109	3
Wind power	0.083	2

Source: Armenia Energy Sector Issues note.

To arrive at our final prioritization ranking, we sum the supply reliability rank and the affordability rank for each investment, and sort them from smallest (highest priority) to largest (lowest priority). Table B.3 shows our final prioritization rank for generation.

Table B.3:   Final Prioritization Ranking for Generation in Armenia

Investment description	Supply reliability rank	Affordability rank	Total
Vorotan HPP	1	1	2
Energy efficiency	1	1	2
Sevan-Hrazdan Cascade	2	1	3
Shnokh HPP	1	2	3
Lori-Berd HPP	1	2	3
Small hydro	2	1	3
Meghri HPP	3	1	4
Pumped storage	1	3	4
Replacement of ANPP	1	3	4
Wind power	2	2	4

Source: Authors.

Transmission

We prioritize transmission investments in Armenia based on supply reliability. Most 220 kV substations have been rehabilitated in Armenia, but all 220 kV overhead lines require rehabilitation. We assess the condition of assets and the importance of investments for ensuring reliable supply based on the following criteria:

- Age: Oldest = highest priority (rank=1)
- Average number of outages: Greatest number of outages = highest priority
- Duration per outage: Longest duration = highest priority

Table B.4 lists the age, average number of outages, and duration per outage for 220 kV and 100 kV transmission lines and their corresponding ranking. We then sum the individual rankings of these three criteria to derive a total supply reliability rank. We use the supply reliability rank as the overall ranking for transmission investments in Armenia because our supply reliability criteria are a proxy for the relative benefits of various transmission investments in Armenia (in terms of number of outages or duration of outages reduced) and because all of the investments are equal in terms of unit costs.

Table B.4:   Supply Reliability and Final Prioritization Ranking of Transmission Investments in Armenia

	Name of line	Age	Rank	Average outages	Rank	Duration/ outage	Rank	Total rank
1	Echmiatsin	54	1	12.5	3	620.1	7	11
2	Shahumyan-2	42	21	10.5	5	239.6	11	37
3	Shinuhayr	45	16	21.5	1	39.4	22	39
4	Lichq	52	4	2	24	115.0	15	43
5	Gougarq-2	39	31	5.5	10	2826.7	2	43
6	Noraduz	52	4	1.5	28	49.0	19	51
7	Bjni	35	38	14.5	2	128.0	13	53
8	Kentron	36	35	3.5	17	4497.3	1	53
9	Sevan	53	3	2	24	15.5	29	56
10	Vardenis	52	4	1.5	28	28.7	24	56
11	Shahumyan-1	54	1	1.5	28	16.0	28	57
12	Gosh	50	9	1	39	203.0	12	60
13	Noyemberyan	46	14	4	13	12.9	33	60
14	Yerevan	46	14	1.5	28	90.0	18	60
15	Gougarq-1	45	16	3	19	17.5	26	61
16	Areg	36	35	7.5	6	42.5	21	62
17	Ninotsminda	10	55	12	4	2095.6	4	63
18	Ashnak-1	28	48	5	11	1469.8	5	64
19	TPP-1	40	25	1	39	2255.5	3	67
20	Shamb	40	25	7	8	8.1	37	70
21	Anoush	35	38	1.5	28	975.0	6	72
22	Mousaler	33	45	4	13	110.5	16	74
23	Megrhi-1	4	58	7	8	367.1	10	76
24	Vayq	52	4	1.5	28	1.3	44	76
25	Sebastia	47	11	1	39	16.5	27	77
26	Norq	44	18	3	19	2.0	42	79
27	Megrhi-2	13	54	4	13	110.5	16	83
28	Tatev-3	38	32	1.5	28	32.0	23	83
29	Lalvar	48	10	1.5	28	0.0	45	83
30	Marash	36	35	2	24	24.5	25	84
31	Tatev-1	38	32	1	39	125.0	14	85
32	Alaverdy	35	38	1	39	428.0	8	85
33	Ani	28	48	7.5	6	12.9	32	86
34	Vorotan-1	52	4	1	39	0.0	45	88
35	Ashnak-2	25	51	1.5	28	374.3	9	88
36	Gyumri	41	24	2.5	22	1.4	43	89
37	Vorotan-2	40	25	1.5	28	2.7	40	93
38	TPP-2	40	25	0.5	50	47.0	20	95
39	Karmir-2	40	25	1	39	14.5	31	95
40	Sipan	35	38	4	13	0.0	45	96

	Name of line	Age	Rank	Average outages	Rank	Duration/ outage	Rank	Total rank
41	Karmir-1	40	25	1	39	11.0	34	98
42	Erebouny	35	38	2	24	5.8	38	100
43	Tatev-2	38	32	1	39	15.5	29	100
44	Goris	31	47	3	19	8.7	36	102
45	Toumanyan-1	47	11	0	51	0.0	45	107
46	Toumanyan-2	47	11	0	51	0.0	45	107
47	Ahar-1	8	56	4.5	12	5.2	39	107
48	Lory	28	48	3.5	17	0.0	45	110
49	Beregovaya	35	38	1.5	28	0.0	45	111
50	Arapnya-1	44	18	0	51	0.0	45	114
51	Arapnya-2	44	18	0	51	0.0	45	114
52	Pambak-1	42	21	0	51	0.0	45	117
53	Pambak-2	42	21	0	51	0.0	45	117
54	Getap	35	38	1	39	2.5	41	118
55	Gagarin	32	46	1	39	10.5	35	120
56	Ahar-2	7	57	2.5	22	0.0	45	124
57	Ghars	23	52	0	51	0.0	45	148
58	Haghtanak	18	53	0	51	0.0	45	149
59	Davit Bek	4	58	0	51	0.0	45	154

Source: Authors and PSRC.

Distribution

We prioritize distribution investments in Armenia by region based on supply reliability. We assess supply reliability based on three criteria:
- Average frequency of outages (number of outages per customer)
- Average duration of outages (minutes of outages per customer)
- Energy not served (minutes of outages per kWh supplied by power stations)

Table B.5 lists the average frequency of outages, average duration of outages, energy not served in each Marz, and the corresponding ranking based on these criteria. We sum the individual rankings of the three criteria to derive a total supply reliability rank. Similar to transmission investments, we use the supply reliability rank as the overall ranking for transmission investments in Armenia because our supply reliability criteria are a proxy for the relative benefits of various distribution investment by Marz in Armenia (in terms of number of outages or duration of outages reduced), and because all of the investments are equal in terms of costs.

Table B.5:   Supply Reliability and Final Prioritization Ranking of Distribution Investments in Armenia

Marz (Region)	Frequency (outages/ customers)	Rank	Duration (minutes/ customer)	Rank	Energy not served (minutes/ kWh supplied)	Rank	Total rank
Syunik	0.0058	1	2.63	1	0.0026	1	3
Tavush	0.0038	2	0.62	4	0.0013	4	10
Vayotz Dzor	0.0028	4	0.59	5	0.0016	2	11
Guegharkunik	0.0025	7	0.66	2	0.0013	3	12
Aragatsotn	0.0027	5	0.63	3	0.0011	6	14
Shirak	0.0036	3	0.57	6	0.0007	7	16
Lori	0.0027	6	0.51	8	0.0012	5	19
Kotayk	0.0024	8	0.53	7	0.0006	9	24
Ararat	0.0014	11	0.38	9	0.0007	8	28
Armavir	0.0019	9	0.36	10	0.0003	10	29
Yerevan	0.0016	10	0.18	11	0.000146	11	32

Source: Authors and PSRC.

The Kyrgyz Republic

The following subsections describe how we prioritize generation and grid development investments in the Kyrgyz Republic.

Generation

We assess supply reliability of generation in the Kyrgyz Republic based on the following parameters:

- Which type of capacity is most needed in order to continue to reliably serve daily load?
- What investments are needed to serve existing demand in the Kyrgyz Republic?
- Does the investment increase fuel supply security in the Kyrgyz Republic?

We rank each of these parameters based on the following factors:

- Role in meeting daily load. We rank the role that each type of generation plays in meeting daily load as follows:
  - Winter baseload = 1
  - Baseload/Peak = 2
  - Not dispatchable = 3 (lowest priority)
- Needed to serve existing demand. We rank whether investments are necessary to serve existing demand as follows:

- – Failure to invest may lead to unforeseen breakdowns in upcoming winter = 1
- – Rehabilitation necessary to meet existing demand in next 3–5 years = 2
- – New capacity necessary to meet growth in demand in next 5–10 years = 3
- Security of fuel supply. We rank security of fuel supply as follows:
  - – Maintains existing level of supply diversity or increases supply diversity using domestic resources = 1
  - – Increases supply diversity OR uses domestic resources = 2

Table B.6 shows how we rank each investment against each of these factors to develop an overall rank for supply reliability. We then assign each investment a corresponding high (1), medium (2), or low (3) rank to allow for comparison with the affordability rank order.

We do not have sufficient data to develop levelized energy costs of investments in the Kyrgyz Republic or to estimate the benefits of each investment. We therefore base our affordability ranking of generation on the overnight cost (US$/MW) of the investment. We rank investments based on overnight costs as follows:

- Less than US$ 100/MW = 1
- US$ 100/MW to US$ 1,000/MW = 2
- Greater than US$ 1,000/MW = 3

Table B.7 shows how we rank generation investments in the Kyrgyz Republic in terms of affordability.

Table B.6: Supply Reliability Rank for Generation in the Kyrgyz Republic

Investment description	Daily load	Importance in serving existing demand	Supply security	Supply reliability rank	Corresponding rank	Comments
Supply of urgently needed equipment and materials for Bishkek CHP and HPPs	1	1	1	3	1	Will provide more reliable supply in upcoming winter
Energy efficiency	1	1	1	3	1	25% of electricity can be saved (2,000 GWh) by reconstructing and modernizing existing energy equipment; 13% savings by means of technical and organizational activities requiring minimal CAPEX
Supply of cables, switches, and other essential spares for Toktogul HPP	2	1	1	4	1	Important to avoid unforeseen breakdown of plant electrical
Rehabilitation of Bishkek CHP (rehabilitation/replacement of 2 steam boilers and selected steam super heaters and transformers)	1	2	2	5	2	Important for adequately serving baseload in upcoming winters
Rehabilitation of Uch-kurgan HPP	2	2	1	5	2	Important for maintaining use of existing capacity and domestic resources
Bishkek CHP	1	3	2	6	2	Needed to ensure adequate winter baseload generation for growing demand
Karakeche TPP	1	3	1	5	2	Needed to ensure adequate winter baseload generation; increases supply diversity using domestic resources

Table B.6: (cont)

Investment description	Daily load	Importance in serving existing demand	Supply security	Supply reliability rank	Corresponding rank	Comments
Kambarata-1	2	3	2	7	3	Size of reservoir increases reliable capacity, but still subject to seasonal water level variability; increases use of domestic resources
Small HPPs	3	3	2	8	3	Increases capacity, but not when needed (during winter peak periods when Toktogul running less)

Source: Authors.

Table B.7: Affordability Rank for Generation in the Kyrgyz Republic

Investment description	Total cost (US$ mln)	US$/kW	Affordability rank
Supply of urgently needed equipment and materials for Bishkek CHP and HPPs[a]	11	85.5	1
Energy efficiency	10	Unknown	1[b]
Supply of cables, switches, and other essential spares for Toktogul HPP[a]	20	25.6	1
Urgent rehabilitation of Bishkek CHP (rehabilitation/replacement of 2 steam boilers and selected steam super heaters and transformers) [a]	8	62.2	1
Rehabilitation of Uch-kurgan HPP	64	439.3	2
Bishkek CHP	150	250.0	2
Karakeche TPP	1350	6097.6	3
Kambarata-1	1700	894.7	2
Small HPPs	255	1432.6	3

Source: Data from utility companies and relevant government agencies.

a. For investments in urgently needed equipment or rehabilitation, we divide total costs by total operational capacity (MW) of the plant based on the assumption that lack of urgent rehabilitation may lead to plant failure in upcoming winter.

b. Energy efficiency assumed to be one of cheapest investments on unit cost basis.

To arrive at our final prioritization ranking, we sum the supply reliability rank and the affordability rank for each investment and sort them from smallest (highest priority) to largest (lowest priority). Table B.8 shows our final prioritization rank for generation.

Table B.8: Final Prioritization Ranking for Generation in the Kyrgyz Republic

Investment description	Supply reliability rank	Affordability rank	Total
Supply of urgently needed equipment and materials for Bishkek CHP and HPPs	1	1	2
Energy efficiency	1	1	2
Supply of cables, switches, and other essential spares for Toktogul HPP	1	1	2
Urgent rehabilitation of Bishkek CHP (rehabilitation/replacement of 2 steam boilers and selected steam super heaters and transformers)	2	1	3
Rehabilitation of Uch-kurgan HPP	2	2	4
Bishkek CHP	2	2	4
Karakeche TPP	2	3	5
Kambarata-1	3	2	5
Small HPPs	3	3	6

Source: Authors.

Transmission

We assess supply reliability of transmission in the Kyrgyz Republic based on the following parameters:

- Is the investment important for improving domestic supply reliability?
- Does the investment increase supply security in the Kyrgyz Republic?

For the first parameter, we rank all investments that improve the reliability of domestic supply as highest priority (Rank = 1). We rank all investments that improve regional interconnections, but do not contribute to domestic supply reliability as lowest (Rank = 3). For the second parameter—supply security, we rank investments as follows:

- Reduces dependence on energy supplied from or transmitted through neighboring countries = 1
- Increases export/import capacity if regional trade arrangements developed = 2
- Does not improve supply security = 3

Table B.9 demonstrates how we rank transmission investments in the Kyrgyz Republic for the above two parameters and how we derive a corresponding supply reliability rank.

Table B.9:   Supply Reliability Rank for Transmission Investments in the Kyrgyz Republic

Investment description	Contributes to domestic supply reliability		Supply security		Total supply reliability rank	Corre-sponding rank
	Rank	Comment	Rank	Comment		
Transmission metering and rehab	1	Metering and rehabilitation essential to reducing losses and ensuring adequate supply	3	Does not contribute to supply security	4	2
Aigultash-Samat 110 kV line	1	Improves reliability of supply in Batken oblast	1	Reduces dependence on energy supplied from Tajikistan	2	1
Datka-Kemin	1	Developing major domestic transmission line crucial to ensuring continued supply reliability for major load centers	1	Disputes with Uzbekistan and Kazakhstan could compromise reliability of lines running through Uzbek and Kazakh territories	2	1
Kemin-Almaty[a]	3	For export purposes only; contributes to supply reliability only if regional trade arrangements well developed	2	Improves supply security only if regional trade arrangements well developed	5	3
CASA 1000 (Datka-Khodjent) [a]	3	For export purposes only; contributes to supply reliability only if regional trade arrangements well developed	2	Improves supply security only if regional trade arrangements well developed	5	3

Source: Authors.

a. Surplus electricity expected to decrease by half by 2022 if no new generation capacity built; surplus available for export likely reduced by more than 50% during dry cycle.

As with generation, we do not have enough data to develop levelized energy cost of investments or to estimate the benefits of each investment. We therefore base our affordability rank on the total cost of the investment, ranking investments as follow:

- Less than US$ 100 million = 1
- US$ 100 to US$ 500 million = 2
- Greater than US$ 500 million = 3

Table B.10 shows how we rank transmission investments in the Kyrgyz Republic in terms of affordability.

Table B.10:   Affordability Rank for Transmission Investments in the Kyrgyz Republic

Investment description	Total cost (US$ mln)	Affordability rank
Transmission metering and rehab	56	1
Aigultash-Samat 110 kV line	12	1
Datka-Kemin	598	3
Kemin-Almaty	140	2
CASA 1000 (Datka-Khodjent)	192	2

Source: Data from utility companies and relevant government agencies.

To arrive at our final prioritization ranking, we sum the supply reliability rank and the affordability rank for each investment, and sort them from smallest (highest priority) to largest (lowest priority). Table B.11 shows our final prioritization rank for transmission investments.

Table B.11:   Final Prioritization Rank for Transmission Investments in the Kyrgyz Republic

Investment description	Supply reliability rank	Affordability rank	Total rank
Transmission metering and rehab	2	1	3
Aigultash-Samat 110 kV line	2	1	3
Datka-Kemin	1	3	4
Kemin-Almaty	3	2	5
CASA 1000 (Datka-Khodjent)	3	2	5

Source: Authors.

Distribution

We rank supply reliability of distribution in the Kyrgyz Republic as follows:

- Reduces technical losses and/or outages = 1
- Reduces commercial losses = 2

Table B.12 demonstrates how we rank distribution investments in the Kyrgyz Republic based on supply reliability.

Table B.12:   Supply Reliability Rank for Distribution Investments in the Kyrgyz Republic

Investment description	Supply reliability rank	Comment
Metering and data acquisition system for the remaining DISCOs	2	Important for reducing commercial losses
Metering and data acquisition system for Severelectro	2	Important for reducing commercial losses
Rehabilitation of distribution assets	1	Important for reducing technical losses and outages

Source: Authors.

We base our affordability rank of distribution investments in the Kyrgyz Republic on the cost of the investment per customer. Table B.13 demonstrates how we rank distribution investments in the Kyrgyz Republic based on supply reliability.

Table B.13   Affordability Rank for Distribution Investments in the Kyrgyz Republic

Investment description	Total cost	Number of customers benefiting	US$ per customer	Affordability rank
Metering and data acquisition system for the remaining DISCOs	24	698,778	$34.35	1
Metering and data acquisition system for Severelectro	36	501,857	$71.73	2
Rehabilitation of distribution assets	190	1,200,635	$158.25	3

Source: Data from utility companies and relevant government agencies.

To arrive at our final prioritization ranking, we sum the supply reliability rank and the affordability rank for each investment and sort them from smallest (highest priority) to largest (lowest priority). Table B.14 shows our final prioritization rank for distribution investments.

Table B.14:   Final Prioritization Rank for Distribution Investments in the Kyrgyz Republic

Investment description	Supply reliability rank	Affordability rank	Total rank
Metering and data acquisition system for the remaining DISCOs	2	1	3
Metering and data acquisition system for Severelectro	2	2	4
Rehabilitation of distribution assets	1	3	4

Source: Authors.

Romania

The following subsections describe how we prioritize generation and transmission investments in Romania We do not have enough data on specific investments planned for the three publicly owned distribution companies and so do not prioritize distribution in Romania.

Generation

We assess supply reliability of generation in Romania based on the following parameters:
- Which type of capacity is most needed in order to continue to reliably serve daily load?
- What investments are needed in order to comply with EU regulations?
- Which investments are likely to be privately financed?

We rank each of these parameters based on the following factors:
- Role in meeting daily load. We rank the role that each type of generation plays in meeting daily load as follows:
    - Baseload = 1
    - Peak = 2
    - Not dispatchable = 3 (lowest priority)
- Compliance with EU regulations. We rank whether an investment is needed in order to comply with EU regulations as follows:
    - Investment promotes compliance with EU regulation = 1
    - Investment does not promote compliance with EU regulation = 2
- Likelihood of private sector participation. We rank the likelihood of private sector participation as follows:
    - Public project = 1
    - Possibility of PPP or private investment with delays = 2
    - Private investment with minimal delays = 3

Justification: Where possible, the Government of Romania is looking to leverage private sector investment to finance generation investment in the power sector. However, a number of these projects are critical for supply reliability in Romania and will need public funding if the private sector does not follow through with the investment. Therefore, understanding where the private sector is and is not likely to finance investments is important for supply reliability of the power system.

We rank each generation investment against each of these factors and sum the rankings to arrive at a total rank for supply reliability. Table B.15 demonstrates our supply reliability ranking for generation investments in Romania.

Table B.15:   Supply Reliability Rank for Investments in Generation in Romania

Investment	Supply reliability		EU regulation		Like-lihood of PSP	Total rank
	Rank	Comment	Rank	Comment		
Environmental upgrade – lignite plants	1	Important for serving baseload	1	Complies with EU Large Combustion Plants Directive	1	3
Energy efficiency Implementation	1	Reduces demand, delays supply demand gap	1	Complies with Directive 2009/28/EC	1	3
Construction of Cernavoda Units 3 and 4	1	Important for serving baseload	1	Helps comply with EU ETS Directive	2	4
HPP rehabilitation/ new capacity	2	Important for serving peak load	2	Does not help comply with EU Directive	1	5
Environmental upgrade or replacement – gas- and oil-fired TPPs	2	Important for serving peak load	1	Complies with EU Large Combustion Plants Directive	2	5
Renewables (primarily WPPs)	3	Not dispatchable	1	Complies with Directive 2009/28/EC	3	7
Environmental upgrade or replacement – hard coal TPPs[a]	1	Important for serving baseload	1	Complies with EU Large Combustion Plants Directive	2	4 [a]

Source: Authors.

a. Ranked lowest because proposed EU regulation to close loss-making hard coal mines would render investments obsolete.

We do not have sufficient data to develop levelized energy cost of investments in Romania. We therefore base our affordability rank of generation on the overnight cost (US$/kW) of the investment. We rank investments based on overnight costs as follows:

- Less than US$ 500/kW = 1
- US$ 500/MW to US$ 1,500/kW = 2
- Greater than US$ 1,500/kW = 3

Table B.16 shows how we rank generation investments in Romania in terms of affordability.

Table B.16:   Affordability Rank for Generation Investments in Romania

Investment	Size of plant (MW)	Cost (mln US$)	Affordability rank
Environmental upgrade – lignite plants	4,178	1,378.3	1
Energy efficiency implementation	Unknown	2,500	1[a]
HPP rehabilitation/new capacity	2,328	823.8	1
Environmental upgrade or replacement – gas- and oil-fired TPPs	2,960	3,654.3	2
Construction of Cernavoda Units 3 & 4	1,310	2,200	3
Renewables (primarily WPPs)	2,496	4,728.4	3
Environmental upgrade or replacement – hard coal TPPs	1,225	Unknown	3[b]

Source: Data from utility companies and relevant government agencies.

a. Energy efficiency assumed to be one of cheapest investments on unit cost basis.

b. Conservatively ranked as least affordable given lack of data and anecdotal evidence demonstrating the high cost of upgrades or replacement needed in order to comply with EU regulations.

Transmission

We do not have sufficient data to develop our own methodology for prioritizing specific transmission investments in Romania. Instead we prioritize investments for transmission based on Transelectrica's Prospective Plan of the Transmission Grid for 2008–2012 with an outlook to 2017. This plan has been approved by ANRE (the regulator). Table B.17 shows generally how Transelectrica's investment plans fulfill the criteria identified in this study for prioritizing investments.

Table B.17:   Which criteria do Transelectrica's key investments fulfill?

Investment	Purpose	Supply reliability			Affordability
		Adequacy	Security	EU regulations	
2009–2010					
• Rehabilitation and modernization of substations • Modernization of command-control protection system in substations • Replacement of transformer units in substations	• Increase supply reliability in key regions • Reduce O&M costs • Facilitate remote control of grid • Create conditions for future interconnections	✓	✓		✓
2010–2016					
• Development of interconnections • Rehabilitation of substations • Grid reinforcement to facilitate integration of RE technologies and Units 3 & 4 of Cernavoda NPP	• Increase connection capacity with neighboring countries • Increase supply reliability • Connect RE and other new capacity to grid	✓	✓	✓	

Source: Authors.

Table B.18 shows how Transelectrica has prioritized specific transmission investments planned for completion between 2010 and 2016 in Romania.

Table B.18:   Prioritization of Specific Transmission Investments in Romania

Type of investment	Specific investment	Planned start year	Planned end year	Cost (mln US$)	Actual project status
Substation	Barbosi 220/110 kV	2008	2010	15.0	Under preparation
Line	Ostrov 220 kV	2008	2011	20.6	Under preparation
Line	LEA 220 kV Cetate-Ostrov	2008	2011	29.5	Under preparation
Substation	Turnu Severin Est 220/110 kV/MT	2009	2011	17.3	Under preparation
Line	LEA 400 kV Gadalin-Suceava	2009	2013	82.6	Under preparation
Substation	LEA 400 kV PdF II-Resita-Timisoara-Arad, including interconnection with Serbia	2009	2014	90.5	Under preparation
Substation	Vilsoara 400 kV	2010	2011	20.6	Under preparation
Substation	Tulcea Vest 400/110 kV/MT	2010	2012	32.9	Under preparation
Substation	Stejaru 220/110 kV/MT	2010	2012	13.0	Under preparation
Substation	Bradu 400/220/110 kV/MT	2010	2012	38.1	Under preparation
Line	Suceava 110 kV/MT	2010	2012	10.3	Under preparation
Substation	LEA 400 kV Suceava-Balti	2011	2013	22.2	Under preparation
Substation	Craiova Nord 220/110 kV/MT	2011	2013	11.5	Prospective
Line	Arad 110 kV/MT	2011	2013	14.8	Prospective
Line	Timisoara 220/110 kV	2012	2013	10.4	Prospective
Substation	Resita 220 kV 110 kV/MT	2013	2014	14.7	Prospective
Substation	Brasov 400/110 kV/MT	2013	2015	38.6	Under preparation
Substation	Domnesti 400/110 kV/MT	2013	2015	30.2	Under preparation
Substation	Pelicanu 400/110 kV/MT	2014	2016	28.5	Prospective

Source: Transelectrica. August 2009. Perspective Plan of the Transmission Grid (PTG) for 2008–2012 with an outlook to 2017. Retrieved on August 7, 2010 from: http://www.transelectrica.ro/PDF/ManagementRET/Plan/PTG%20Perspective%20Plan%202008-2012%20and%202017.pdf.

Serbia

The following subsections describe how we prioritize generation and transmission investments in Serbia. We do not have sufficient data on the specific investments planned for distribution and so do not prioritize distribution investment in Serbia.

Generation

We do not have enough data to develop our own methodology for prioritizing specific generation investments in Serbia. Instead we prioritize investments for generation based on EPS' planned investments for 2008–2015. These investments generally reflect the government's strategic objectives, ranked as follows:

- Complying with EU requirements = 1 (highest priority)
- Rehabilitating existing capacity = 2
- Building new capacity = 3

Table B.19 shows how EPS' investment plans—listed based on planned start year—generally reflect these objectives.

Table B.19:   EPS' Investment Plans for Generation in Serbia

Investment	Units	Cost (mln US$)	Size (MW)	Planned years of implementation	Strategic objective ranking
Reconstruction or replacement of the existing electrostatic precipitators on TPP units	TPP Kolubara A: unit A5	6	32	2009	1
Reconstruction of the ash and slag transport and disposal system – introduction of the new technology	TPP Kolubara A: unit A5	6	32	2009	1
Primary measures for the reduction on NO emissions from TPP units	TPP Nikola Tesla: A3-A6, B1-B2 TPP Kostolac A & B: A1-A2, B1-B2 TPP Nikola Tesla B: B2 TPP Morava	116	3603	2009-2015	1

Investment	Units	Cost (mln US$)	Size (MW)	Planned years of implemen- tation	Strategic objective ranking
Reconstruction or replacement of the existing electrostatic precipitators on TPP units	TPP Nikola Tesla B: B2 TPP Morava TPP Kostolac B: B1-B2	30	1442	2010	1
Revitalization	HPP Vlasinske Hidroelektrane	56	128	2010-2011	2
Reconstruction of the ash and slag transport and disposal system - introduction of the new technology	TPP Nikola Tesla A: unit A3-A6	56	1231	2010-2012	1
Flue gas desulphurization on the TPP units (FGD)	TPP Nikola Tesla B	292	1240	2010-2013	1
New construction	Completion of TPP Kolubara B construction CHP Novi Sad reconstruction	766 390	700 478	2010-2015	3
New construction	Construction of the new TPP Nikola Tesla B3	1211	700	2011-2015	3
Flue gas desulphurization on the TPP units (FGD)	TPP Nikola Tesla A - unit A3-A6	278	1231	unknown	1
New construction	HPP Gornja Drina (Sutjeska, Buk Bijela, Foca and Paunci) PSHPP Bistrica	606 436	280 680	unknown	3

Source: Authors and EPS.

Transmission

We do not have sufficient data to develop our own methodology for prioritizing specific transmission investments in Serbia. Instead we prioritize investments for transmission based on EMS' planned investments for 2008–2015. We use EMS' investment plans for our prioritization because:

- EMS knows the technical constraints of the transmission network. As the transmission system operator, EMS is best placed to coordinate investments while maintaining system reliability while rehabilitating and expanding the transmission grid.

- EMS knows the condition of specific assets. In general, 46 percent of substations are in poor condition and 28 percent of overhead lines are in poor condition, making substations generally a more important investment from a reliability standpoint. However, EMS' investment plans reflect knowledge of the condition of specific assets, so they show a mix of when specific substations and overhead lines need to be revitalized or upgraded.

Table B.20 shows EMS' investment plans for transmission in Serbia from 2008–2015.

Table B.20:   EMS' Plans for Transmission Investments in Serbia for 2008–2015

Type of investment	Specific investment	Cost (mln US$)	Years to complete
Substation revitalization	Revitalization of the facility close to power plants	24.5	6
	HV equipment and 400kV replacement	41.8	6
	SS 110/35/10 kV Beograd 1	8.4	5
	Revitalization of structural parts	7.1	5
Substation upgrade	HV equipment and 220kV replacement	11.1	4
	Works on SS and SG defined by Annual plans	3.1	2
	Procurement of equipment and materials for construction investment	8.4	6
OHL revitalization	220kV OHL-reconstruction and switching to 400kV	83.6	6
	220 kV OHL revitalization	11.2	6
	110 kV OHL revitalization	11.1	6
OHL upgrade	110 kV OHL Beograd 5 – Stara Pazova	1.0	2
	Works on 110 kV OHL defined by Annual plans	8.4	6
OHL new construction	OHL 400 kV Srbija - Rumunija	34.8	6
	110 kV OHL Majdanpek 2 – Mosna	3.3	2
Technical management and telecommunication	Technical management system	3.6	6
	Telecommunication equipment	22.4	6
Substation revitalization	SS 110/35 kV Požarevac	0.9	1
	SG 110 kV Pančevo 1	2.5	3
Substation upgrade	SS 400/220/110 kV Smederevo 3	11.1	2
	SS 400/220/110 kV Kraljevo 3	9.1	3
OHL upgrade	110kV OHL Novi Sad3-(Novi Sad 7)-Novi Sad5	0.9	1

Type of investment	Specific investment	Cost (mln US$)	Years to complete
OHL new construction	OHL 400 kV Kraljevo 3 – Kragujevac 2	13.9	3
	OHL 110 kV Kraljevo 3 – Novi Pazar 2	7.0	3
	110 kV OHL Bela Crkva – Veliko Gradište	3.9	3
	110 kV OHL Guča – Ivanjica	3.6	2
	110 kV OHL HE Zvornik – Loznica	4.5	2
	Construction of new 110 kV OHL to increase supply security	27.9	5
	110 kV OHL HE Đerdap 2 – Mosna	4.3	2
Substation upgrade	SS 400/110 kV Bor 2	4.2	1

Source: Authors and EMS.

Ukraine

The following subsections describe how we prioritize generation and grid development investments in Ukraine. For both categories of investments, we only consider projects that are expected to begin before 2015.

Generation

We assess supply reliability of generation in Ukraine based on the following parameters:

• Which type of capacity is most needed in order to continue to reliably serve daily load?
• How many customers will be impacted if an investment does not occur?
• Does the investment increase fuel supply security in Ukraine?

We rank each of these parameters based on the following factors:

• Role in meeting daily load. We rank the role that each type of generation plays in meeting daily load as follows:
    – Peak = 1 (highest priority)
    – Baseload = 2
    – Not dispatchable = 3 (lowest priority)

Justification: Ukraine does not have sufficient peak capacity and so has used TPP load-shedding to fill the gap. The operation of TPPs in a manner for which they were not designed has contributed to the premature deterioration of these assets. To continue to meet peak load and to further prevent asset deterioration, we rank peak load highest.

- Sector impact. We rank investments based on a rough assessment of the number of customers that would be affected as follows:
  - System wide impact = 1
  - Large regional impact = 2
  - Smaller regional impact = 3
- Security of fuel supply. We rank security of fuel supply as follows:
  - Uses domestic resources = 1
  - Uses imported fuel supply with potential for development of domestic fuel supply = 2
  - Uses imported fuel supply with no potential for development of domestic fuel supply = 3

We rank each generation investment against each of these factors and sum the rankings to arrive at a total rank for supply reliability. We then assign each investment a corresponding high (1), medium (2), or low (3) rank to allow for comparison with the affordability rank order. Table B.21 demonstrates how this is done for the generation investments being considered in Ukraine.

Table B.21:   Supply Reliability Rank for Investments in Generation in Ukraine

Investment description	Role in meeting daily load	Sector impact	Security of fuel supply rank	Total supply reliability	Corre- sponding rank
HPP rehabilitation	1	1	1	3	1
TPP rehabilitation, retrofit, and replacement	2	1	1	4	1
Service life extension of NPPs	2	1	2	5	1
Completion of Khmelnitski 3 & 4 NPP	2	1	2	5	1
Construction of 4000 MW NPP (including Khmelnitski 5 & 6)	2	1	2	5	1
Construction of 4 new CCGTs in Crimea	1	2	3	6	2
CHPs at Kyivenergo (Rehab CHPs 5 & 6; re-equipment of switchgears)	3	2	3	8	2
WPP new construction	3	3	1	7	2
Kharkiv CHP-5 (CCGT construction)	3	3	3	9	3
Rehabilitation at other CHPs	3	3	3	9	3

Source: Authors.

We assess affordability as the impact on end-user tariffs by comparing the levelized cost (LEC) of each investment. Levelized costs are ranked as follows:

- Less than US$ 0.04/kWh = 1
- Between US$ 0.04/kWh and US$ 0.08/kWh = 2
- Greater than US$ 0.08/kWh = 3

Table B.22 shows the levelized costs of each investment and the investment's corresponding affordability rank.

Table B.22:   Affordability Rank for Investments in Generation in Ukraine

Investment description	Levelized energy cost (US$/kWh)	Corresponding rank
HPP rehabilitation[a]	0.0853	2
TPP rehabilitation, retrofit, and replacement[b]	0.011	1
Service life extension of NPPs[c]	0.016-0.028	1
Completion of Khmelnitski 3 & 4 NPP[d]	0.109	3
Construction of 4000 MW NPP (incl. Khmelnitski 5 & 6) [d]	0.109	3
Construction of 4 new CCGTs in Crimea[d]	0.064	2
CHPs at Kyivenergo (Rehab CHPs 5 & 6; re-equipment of switchgears)	No data	2
WPP new construction[d]	0.08	2
Kharkiv CHP-5 (CCGT construction)[d]	0.064	2
Rehabilitation at other CHPs	No data	2

a. Author calculation based on capital costs for 3rd Stage of UHE Rehabilitation Program.

b. IMEPower. August 2008. Ukraine TPP Rehabilitation: Assessment of Needs, Costs and Benefits. Prepared for World Bank.

c. Author calculations based on capital cost data from: IAEA. "Cost Drivers for Assessment of Nuclear Power Plant Life Extension." September 2002.

d. Based on World Bank estimates of levelized energy costs for a new nuclear plant in Armenia, contained in a World Bank Armenia Energy Sector Issues note (unpublished).

To arrive at our final prioritization ranking, we sum the supply reliability rank and the affordability rank for each investment and sort them from smallest (highest priority) to largest (lowest priority). Table B.23 shows our final prioritization rank for generation.

Table B.23: Final Prioritization Rank for Generation in Ukraine

Investment description	Supply reliability rank	Affordability rank	Total
TPP rehabilitation, retrofit, and replacement	1	1	2
Service life extension of NPPs	1	1	2
HPP rehabilitation	1	2	3
Completion of Khmelnitski 3 & 4 NPP	1	3	4
Construction of 4000 MW NPP (including Khmelnitski 5 & 6)	1	3	4
Construction of 4 new CCGTs in Crimea	2	2	4
CHPs at Kyivenergo (Rehab CHPs 5 & 6; re-equipment of switchgears)	2	2	4
WPP new construction	2	2	4
Kharkiv CHP-5 (CCGT construction)	3	2	5
Rehabilitation at other CHPs	3	2	5

Source: Authors.

Transmission

We assess supply reliability of transmission based on the following parameters:

- What is the current condition of the asset(s)?
- How many customers will be affected if an investment does not occur?
- Will the investment improve supply security by increasing Ukraine's import potential?

For all transmission investments, except Union for the Coordination of Electricity Transmission (UCTE) development, we use data on "total benefits" as a proxy measure of both the condition of the assets and how many customers will be impacted by the investment. Total benefits are derived as follows:

Total benefits = Reduction in Energy Not Served (ENS) + Avoided Losses. We then sort total benefits from highest to lowest and rank them as follows:

- Total benefits > 4 = 1 (highest priority)
- 4 > total benefits > 0.25 = 2
- Total benefits < 0.25 = 3 (lowest priority)

We rank the two grid development investments for which we lack total benefit data as lowest priority for the following reasons:

- UCTE development is important from a supply security perspective, but does not necessarily contribute directly to improvement of the domestic grid. We assume, therefore, that UCTE contributes less to supply reliability in terms of reducing ENS and avoiding losses than other transmission investments.
- Distribution rehabilitation contributes to the reduction in ENS and avoided losses, but any single distribution investment generally affects a smaller number of customers than any single transmission investments.

We also consider security of supply in our assessment of supply reliability for grid development investments. We rank UCTE development as the highest priority because it increases import potential. We rank all other grid development investments as lowest priority as they do not affect supply security.

Table B.24 shows how investments are ranked in terms of total benefits and supply security and how a final rank is developed for supply reliability.

Table B.24:   Supply Reliability Rank for Grid Development in Ukraine

Investment description	Reduction in ENS	Avoided losses	Total benefits	Benefit rank	Supply security rank	Total supply reliability rank	Corresponding rank
330/110 kV substation "Zapadnaya Kiev"	0.67	4.31	4.98	1	3	4	1
Stabilization of Crimea Power Grid - Phase II	0.16	5.39	5.55	1	3	4	1
750 kV Line Zaporizska NPP-Kakhovska	0	6.16	6.16	1	3	4	1
330 kV Line Arctyz-Novo-Odessa[a]	0.09	0.15	0.25	1	3	4	1
UCTE Development	No data			3	1	4	1
330 kV Lutsk Pivnichna-Ternopol	0.91	0.93	1.83	2	3	5	2
Voltage level normalization	0	0.41	0.41	2	3	5	2
330 kV Zarya-Mirna #2	0	0.06	0.06	3	3	6	3

Source: Authors and Decon/KfW.

a. This investment is considered necessary to satisfy N-1 criteria and so is ranked 1 even though benefits appear low.

We use a benefit-cost ratio as a proxy measure of the affordability of grid development investments. We rank benefit-cost ratios as follows:

- Benefit-cost ratio > 1 = 1
- 1 > benefit-cost ratio > 0.25 = 2
- Benefit-cost ratio < 0.25 = 3

We do not have benefit-cost data on UCTE development. On a total cost basis, UCTE development is the most expensive of the grid development investments, second only to rehabilitation of the entire distribution network. Based on the measure of total benefits we have used—assessing total benefits in terms of supply reliability, but not supply security—UCTE development also ranks relatively low. In order to not overestimate the value of UCTE development in terms of affordability, we conservatively rank UCTE development as lowest priority (Rank = 3).

Table B.25 shows the benefit-cost ratio of each investment and the corresponding affordability rank.

Table B.25:   Affordability Rank for Grid Development in Ukraine

Investment description	Benefit-cost ratio	Affordability rank
330/110 kV substation "Zapadnaya Kiev"	1.58	1
Stabilization of Crimea Power Grid - Phase II	0.67	2
750 kV Line Zaporizska NPP-Kakhovska	0.26	2
330 kV Line Arctyz-Novo-Odessa*	0.04	3
UCTE Development		3
330 kV Lutsk Pivnichna-Ternopol	0.28	2
Voltage level normalization	0.12	3
330 kV Zarya-Mirna #2	0.16	3

Source: Authors and Decon/KfW.

To arrive at our final prioritization ranking, we sum the supply reliability rank and the affordability rank for each investment and sort them from smallest (highest priority) to largest (lowest priority). Table B.26 shows our final prioritization rank for grid development.

Table B.26:   Final Prioritization for Grid Development in Ukraine

Investment description	Supply reliability rank	Affordability rank	Total
330/110 kV substation "Zapadnaya Kiev"	1	1	2
Stabilization of Crimea Power Grid - Phase II	1	2	3
750 kV Line Zaporizska NPP-Kakhovska	1	2	3
330 kV Line Arctyz-Novo-Odessa*	1	3	4
UCTE Development	1	3	4
330 kV Lutsk Pivnichna-Ternopol	2	2	4
Voltage level normalization	2	3	5
330 kV Zarya-Mirna #2	3	3	6

Source: Authors.

ECO-AUDIT
Environmental Benefits Statement

The World Bank is committed to preserving endangered forests and natural resources. The Office of the Publisher has chosen to print **Outage** on recycled paper with 50 percent postconsumer fiber in accordance with the recommended standards for paper usage set by the Green Press Initiative, a nonprofit program supporting publishers in using fiber that is not sourced from endangered forests. For more information, visit www.greenpressinitiative.org.

Saved:
- 5 trees
- 2 million BTU's of total energy
- 468 lbs. of CO_2 equivalent of greenhouse gases
- 2,254 gallons of waste water
- 137 lbs. of solid waste

green
press
INITIATIVE

www.ingramcontent.com/pod-product-compliance
Lightning Source LLC
Chambersburg PA
CBHW070408200326
41518CB00011B/2113